Energy Efficiency Improvements in Smart Grid Components

Energy Efficiency Improvements in Smart Grid Components

Editor

Jens Omdahl

scitus
academics

Energy Efficiency Improvements in Smart Grid Components
Edited by **Jens Omdahl**

ISBN: 978-1-68117-298-9
Library of Congress Control Number: 2016937483

SCITUS Academics LLC,
www.scitusacademics.com
Box No. 4766, 616 Corporate Way,
Suite 2, Valley Cottage,
NY 10989

Notice

Printed in United States of America on Acid Free Paper

Preface

Smart grid usually refers to a class of technology that people use to deliver electricity to utilities through computer - based remote control and automation in the 21st century. These systems are facilitated through two - way communication and computer processing technology, which has been used in other industries for decades. They start to be used in electricity networks, from power plants and wind farms to home and business electricity consumers. They deliver many advantages to utilities and consumers-mainly seen in major energy efficiency improvements in the electricity grid and in the homes and offices of energy users. Electronic power conditioning and control of electricity production and distribution are important elements of the intelligent grid. In order to allow appropriate billing according to the (highly variable) level of consumption of different users, it was necessary to measure the electricity consumption per user. Due to the limited data collection and processing capacity during the period of grid growth, fixed-tariff requirements and dual-tariff requirements where night power was charged at a lower rate than day power were commonly established. The reason for double-tariff arrangements was the lower demand for night time. Dual tariffs enabled low - cost night - time electricity to be used in applications such as the maintenance of 'heat banks' to 'smooth out' the daily demand and to reduce the number of turbines to be switched off overnight, thus increasing its use and profitability of distribution and transmission facilities. This book, Energy Efficiency Improvements in Smart Grid Components, is intended for professors and technicians working in universities, academic institutions, utilities and industry that need to improve their ideas and acquire new vital information about the development of energy productivity in the power grid. It involves concepts for optimizing the power system, embracing value proposals, protecting the power system, etc.

Table of Contents

1

CHAPTER

ENERGY EFFICIENCY IN SMART GRID: A PROSPECTIVE STUDY ON ENERGY MANAGEMENT SYSTEMS

Hermes José Loschi, Julio Leon, Yuzo Iano, Ernesto Ruppert Filho, Fabrizzio Daibert Conte, Telmo Cardoso Lustosa, Priscila Oliveira Freitas

Department of Communications, Faculty of Electrical and Computer Engineering, University of Campinas (UNICAMP), Campinas/SP, Brazil

ABSTRACT

The term Smart Grid has become a term to represent the benefits of a smart and sophisticated electrical grid, which can meet various social expectations related to sustainability and energy efficiency. The Smart Grid promises to enable a better power management for energy utilities and consumers, to provide the ability to integrate the power grid, to support the development of micro grids, and to involve citizens in energy management with higher levels of responsibility. However, this context comes with potential pitfalls, such as vulnerabilities to cyber-security and privacy risks. In this article, a prospective study about energy management, and exploring critical issues of modeling of energy management systems in a context Smart. Grid is presented along with background of energy management systems. An analysis of the demand response condition is also presented. Finally, the advantages and disadvantages of the implementation of energy management systems, and a comparison with the Brazilian electricity system are discussed.

Keywords: Smart Grid, Management, Cloud Computing, Energy, Efficiency

1. INTRODUCTION

Electric energy is essential to increase productivity and ensure a high quality of life; therefore, the relationship between electric power and economic growth is crucial. However, the consequence of the current worldwide economic growth and electricity demand is the depletion of energy resources. An essential and effective way to prevent the depletion of resources and promote economic growth at the same time is the application of the concept of energy efficiency through energy management systems, this is being the basic principle of the Smart Grid [1] .

With the development of the Smart Grid, especially in the distribution grid, and with the possibility of load modeling, control over the peaks of energy demand becomes vital. The peaks of demand are serious problems and present themselves in the electrical system. The demand management in residential, commercial and industrial sectors can play an important role in reducing peak demand, reducing stress, overhead transmission and distribution lines. In many countries, there are various demand response programs, implemented for industrial and commercial loads [2] .

There are few demand response (DR) programs in use for energy management in the residential sector. Direct load restriction is the most popular method used to reduce peak demand. However, when using direct control/restriction load, the consumer comfort can be compromised. In contrast, reduction of peak demand through load displacement can benefit consumers and energy utilities [2] .

Peak demand of energy has caused adverse effects to the reliability and stability of the power system during recent decades. Reducing peak demand can reduce the risk of faults on transmission and distribution grid, consequently, the risk of interruptions. Demand response is one way to deal with peak events and avoid overload on the grid by providing the necessary flexibility through load displacement [2] [3] .

Multiple energy management systems use the concepts of demand response; however, these DR systems are not broadly implemented due to complexity of the automation involved, particularly in industrial and commercial buildings. Although the residential sector is responsible for a significant amount of electrical energy demand, few DR programs are currently used in the residential field. Direct Load Control (DLC) is the method most used by energy utilities in Brazil to manage peak demand in the residential sector, where consumer loads are adjusted in time during peak network events [4] [5] .

An energy management system that can automatically switch the operation of appliances during peak hours, can be used to management of peak demand without compromising the quality of supply to the consumer [6] .

Up recognizing the very different perspectives and priorities of individuals and organizations involved in the electrical system, the authors propose a broad analysis through a prospective study to examine the different ways for the Smart Grid meeting the demands and developments of society. This study focuses on comparing the development of "Smart Grid" in different regions of the world and demonstrating the commitment of these countries to change the social and political contexts and expectations, which often are shaped by specific regions, goals and available resources.

2. SMART GRID TECHNOLOGIES

The term Smart Grid means more than a single technology or even a clear set of individual technologies. Is an "umbrella" term under which various technologies of electric power systems are considered, both in hardware and software. For some people, Smart Grid is characterized primarily as the addition of an information and communication technology (ICT), superimposed in a way on existing infrastructure. For others, Smart Grid represents the installation of new transmission lines, meters, and renewable energy generation [7] .

However, in order for both conditions to comply, first it is necessary to understand the legacy electric systems worldwide. The current dominant infrastructure of electric power systems involves four basic elements:

- Generation: Electric power is generated in large-scale power plants;
- Transmission: High-voltage electrical energy is transported from the plant to substations closer to consumers;
- Distribution: Low voltage energy is distributed from substations to residences and commercial buildings;
- Consumer: Electricity used for consumer devices such as refrigerators, computers, lights, pumps and other devices used by residential, commercial, and industrial devices.

The main mechanism in power generation by current systems depends on the heat produced by burning fossil fuels, division of atoms in nuclear energy, or from the hydroelectric stations water movement. Except for solar cells, almost all other forms of power generation, including the burning of fossil fuels, nuclear, biomass, hydro, wind, concentrated solar, cogeneration, and need driving a turbine to produce electricity [7] [8] .

The generation usually produces electricity with relatively low voltages ranging from 2 to 30 kilovolts (kV), depending on the size of the unit. Since electricity is generated, its tension is amplified before transmission. A critical step between the generation of electricity and long-distance transmission involves a transformer to increase the voltage. Often, the generation of electric energy occurs far from the places where the electricity is needed, making the long distances of high voltage transmission lines, a crucial part of the electrical system. The long-distance transmission voltage varies from 115 to 120 kilovolts (KV), so the transformer plays a crucial role in increasing the voltage for transmission [7] [8] .

The high-voltage transmission lines carry electricity from generating plants to local substations, where the energy is "left over" for a lower voltage and then sent through electric energy distribution networks for local users, including the industrial, commercial, and residential consumers. From the substation, the electrical energy is distributed locally within a community to individual buildings and houses. The voltage is usually reduced at the point of use, to the standard voltage of that region, which varies in different countries (with most consumers receiving 110 - 120 V in the United States and 220 - 240 V in Europe) and with the requirements of electric power use [7] [8] .

MAIN TECHNOLOGIES

The term Smart Grid represents the integration of digital technologies, sensors, and ICTs to empower and make the management on the use of electricity more reliable and efficient. Smart Grid includes technologies for the consumer (with which consumers interact) and the grid (transmission and distribution that are less visible to consumers). The Smart Grid technologies also include hardware and software [7] [8] .

One of the definitions of the term Smart Grid is the integration of various technologies, products and services, from the generation, transmission, and distribution; using advanced communication and control technologies. Figure 1 illustrates this concept.

Table 1 presents the main technologies proposals for Smart Grid and their definitions.

In addition to the above-mentioned Smart Grid technology, a more holistic approach to energy management systems is presented in the next section. A major challenge with current systems is the limited mechanisms for coordination and communication between the management of the different parts of the system. The management of the current systems' transmission and distribution are carried out through separate activities that occur in different parts of the network [8] .

3. ENERGY MANAGEMENT SYSTEMS IN SMART GRID

The vast majority of the energy management systems just consider the monitoring and data statistics of energy consumption of consumer electronics. For these systems, manual actuation is necessary in each device to reduce energy consumption. However, the Smart Grid technologies require management systems to be smarter and able to respond to demands related to the charge control, energy management, and timing systems with micro grids [9] .

Some architectural solutions for energy management systems are being integrated into a concept of Smart Grid; these are presented blow. Chang-Sci Choi develops an architecture using AMI solutions for an energy management system entitled EMM (Energy Monitoring and Management), interoperating this system with the Smart Grid. Among the assumptions adopted, the most important is the division of the stream of operation and settings of residences and buildings [9] .

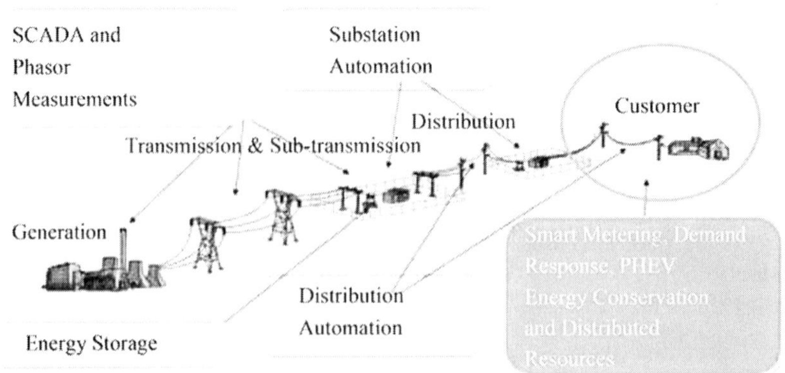

Figure 1. Concept for control and communication systems in Smart [9] .

Table 1. Main technologies proposals for Smart Grid [7] .

Phases	Technologies	Definition
Generation	Inverters "Grid Tie"	Use of "Grid Tie" inverter, connecting to mains, e.g. photovoltaic systems
	Synchrophasors	Measurement of Synchronized Phasor, i.e. sinusoidal measurements of AC magnitudes, and synchronized in time expressed in phasors. To determine useful information about grid performance
Transmission	FACTS (Flexible Alternate Current Transmission Systems) para HVAC (High Voltage Alternating Current) e HVDC (High Voltage Direct Current)	AC or DC (High Voltage ou Low Voltage) voltage, transmission from generation to distribution
	Static Deviation/Compensator VAR (Volt-amper reactive)	Used for static VAR compensation, have mutual inductance lines, consuming reactive power
Test and modeling of transmission	Software for Analysis of Transmitted Power	Full suite of tools to create, configure, customize and manage power transmission systems
Transmission in cases of distribution of HVDC and generation by renewable sources	Inverters and rectifiers	Used for conversion AC-DC and DC-AC
Transmission and distribution	Substation automation	Applications of automation: Control, timing, voltage transfer, reduction and detection of loads
Transmission, distribution and substation	Relays and circuit breakers	The relay is actuated to detect any failure in the power system, disarming the breakers of the circuits, avoiding damage to the equipment
Distribution (Distribution management system-DMS)	Fault location for distribution System.	Fault locators are devices and software, usually installed in substations to identify possible fault events, calculating the distance from the point of the failure to the source monitored identified in distribution system
Distribution	AMI (Advanced metering Infraestrutura)	AMI provides the communication with the smart meters and other power management devices
Distribution, information management.	Distribution automation	The system consists of equipment's, communication infrastructure and information technology, which are used as intelligence distribution system
	Management system of measurement data	Used for complex data collection processes measured for multiple data recording technologies
	RTU (Remote Terminal Unit)	The function of the RTU is the remote location of the system SCADA, for monitoring and control of the necessary equipment
Information management	SCADA (Supervisory Control and Data Acquisition)	Computational system that monitors and controls infrastructure and industrial plant processes based installation
	EMS (Energy Management System)	The EMS are tools used for operation of electrical grid, to monitor, control and optimize the performance of systems for generation, transmission and distribution
Information management and consumer economics	Smart Meters	The smart meter registers the range of the electric energy consumption

Continued

Information management: HAN (Home are network) and LCM (Load Control Modules).	PCT (Program Communication Thermostat)	Components of the control system, to detect the temperature of a system, controlling the same in the desired set point.
Residential Consumption	Load control receiver	Are devices used for load control, directly or indirectly, through voltage circuits, such as air conditioning thermostat.
Consumption of the generation, distribution and transmission	Short circuit current limiter	Current limitation is the practice in electrical or electronic circuits, the imposition of an upper limit to the current that can be delivered, avoiding damage to the equipment
Transformers generation, distribution and transmission	OLTC (On-Load Tap-Changer)	The OLTC is used to change the transformer voltage ratio, without interrupting the load service

The sizing system EMM first considered complex apartments and residential environments popular in South Korea, with advanced infrastructure of home networks and several features, such as: electricity, gas, water, and cogeneration. Figure 2 shows the architecture that is installed via the internet providing accessibility, mobility, and interoperation with other systems. Each EMM installed performs communication with smart watches "Smart Meters", and interface with networked home appliances. To report the current status of the house, the energy measurement system sends data in real time via the Internet [9] .

For commercial buildings and industrial plants, the author proposes the BEMS (Building Energy Management System). However, these systems and its communication protocols depend on the market and corporate strategies at the time of construction, and its concepts of energy efficiency and energy management, which adopt generic and growing open communication standards and protocols, such as: BACnet, LonWorks, Modbus, KNX, WLAN, Zigbee, SNMP, IEC61850, DNP3, etc. To overcome these constraints, the author proposes the development of 2 CCL (Common Communication Layer) protocols. The first one is called BAS CCL, and the other one, EMM CCL. Figure 3 illustrates the configuration of EMM for heterogeneous integration BEMS in each building [9] .

In both systems, the server data is analyzed by software, providing information about weather and conditions of operation of the entire plant. This information can be broadly used and incorporated into a maintenance planning, among other applications [9] [10] .

Levels of data analyzed by the software can be broadly classified as:

- Level 1, Management: Supervision of computers, servers and services data management servers;

- Level 2, Automation: Smart G/W (gateway);
- Level 3, Installation: Sensors, actuators and controllers.

The software operates considering a data group, composed of service server and data management that are installed on the building control center, as illustrated in Figure 4. The Smart Gateway is installed inside the building, collecting the data of the electric power consumption through intelligent devices such as meters, sensors and actuators, etc., that are engaged in the construction of an operating system, including electricity control systems, HVAC/HVDC, lighting systems, etc. [10] .

This architecture allows the system to provide multiple functions for energy saving control, such as: control of maximum demand of energy, weather-based light and dimming control by means of sensors. The information generated by the system makes it possible to provide the function of energy efficiency through predictive analysis of the trend of energy consumption, comparing with the data in real time, through a comparative analysis with similar facilities. Another option is the provision of services, such as load shifting adapting to pricing/rates through Smart Grid function; connection with renewable energy sources; and energy exchange function. In addition, it allows a selection of ideal tariff systems appropriate to the construction standard and energy consumption of the installation [10] .

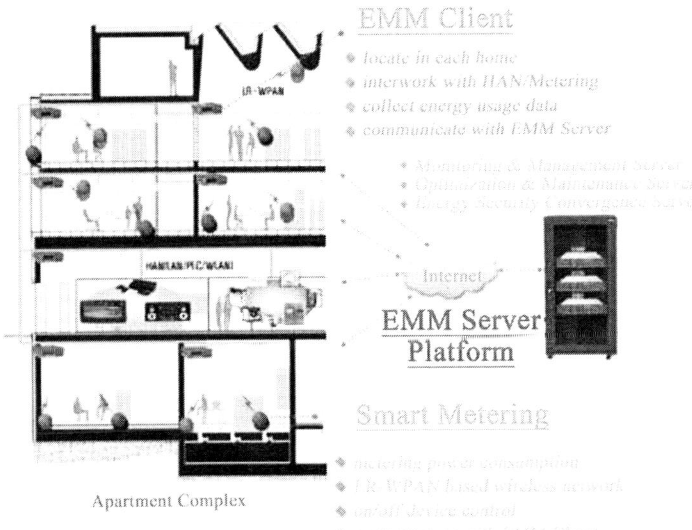

Figure 2. Systems configuration of EMM for residence [9] .

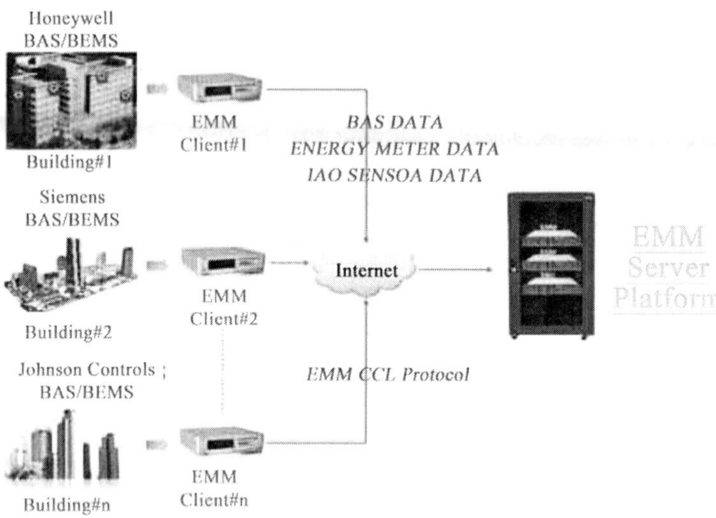

Figure 3. Systems configuration of EMM for integration with BEMS [9] .

4. ANALYSIS OF THE CONDITION OF DEMAND RESPONSE

To meet peak demand, high-cost generating stations are required. Adding more generation was the strategy used in the past to meet the demand of electricity. Currently, the energy utilities have given more attention to demand management in order to reduce peak demand. Using the Smart Grid concept means more than a single technology or even a clear set of individual technologies for this [2] .

Demand response (DR) is a key concept in energy demand management, which helps to reduce peak demand in critical situations. DR is defined as the changes in the use of electricity, for end consumers, of their normal consumption patterns in response to changes in the price of electrical energy over time, or the incentive payments intended to induce a better use of electricity at peak hours [11] .

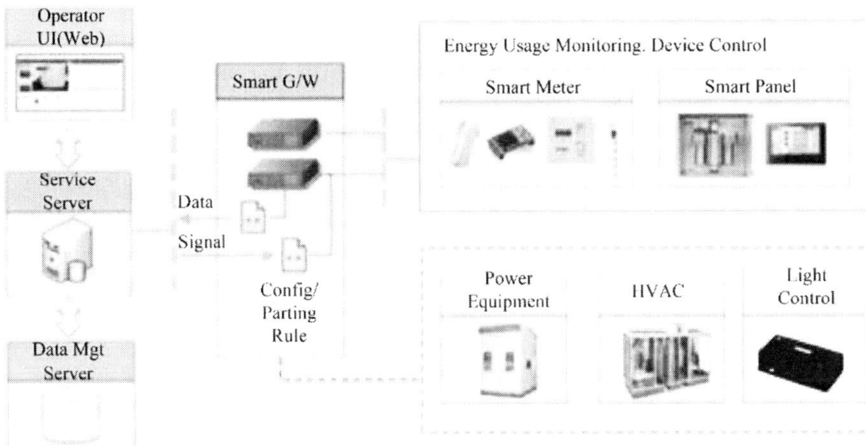

Figure 4. Performance software architecture [10] .

Load management is defined as a set of objectives which aim to control directly or indirectly and/or modify patterns of electricity consumption of various consumers, aiming to reduce peak demand. This control and modification enable the supply system to meet the demand, making better use of its available generation and transmission capacity [12] .

For leveling the peaks of demand, three common strategies for load management are used: "Peak Clipping", "Load Shifting" and "Valley Filling", as illustrated in Figure 5.

"Peak Clipping": Load reduction for short peaks and periods of use, usually performed by the direct load control. In this method, the energy utilities disconnect the consumer when there is a critical situation. This direct control can be used to reduce capacity requirements, operational costs and dependence on fossil fuel generation [12] .

"Valley Filling": Creation of loads during the peak period. This helps to reduce the average price of electricity. One of the methods used in industrial production, which uses the loads generated by fossil fuels [12] .

"Load Shifting": Moves the peak loads for other periods of time without necessarily changing the global consumption. This method combines the benefit of "Peak Clipping" and "Valley Filling" moving existing loads during off-peak hours [12] .

In programs of DR, electricity consumers play an important role in the reduction of peak demand during peak hours. Consumers can move their loads and thus help the energy utilities to prevent failures and blackouts in the electrical system, reducing the probability of stress conditions of the system. Improve energy security through the DR increases productivity and customer satisfaction. The DR also eliminates the need for high-cost generators and eventually reduces the cost of electricity [2] .

In order to inform consumers with real-time data, there must be a communication link between the energy utilities and consumers. Consumers must be able to measure their electrical energy demand, in real time, in order to act for demand response events. Advanced metering infrastructure implementations (AMI) and other technologies allows the user to measure the real-time energy demand and further enhance the use of resources of DR in daily operation [13] .

Therefore, it is evident that there is a need for an automatic energy management system in DR programs, which will provide more flexibility consumers.

5. CLOUD COMPUTING SOLUTIONS IN SMART GRID

The term cloud computing has many definitions; in scenarios as Smart Grid, it is defined as a model for enabling convenient, on-demand network access to a shared pool of configurable computing resources (e.g. networks, servers, storage, applications, and services) that can be rapidly provisioned and released with minimal management effort or service provider interaction. The characteristics of cloud computing include on-demand service, ubiquitous network access, location independent resource pooling, rapid elasticity, and measured service.

Cloud computing is not a universal solution. It has strengths and weaknesses, and understanding them is the key to making a decision about whether it is right for a particular application [14] . The main advantages of cloud computing are: Economy of scale: sharing of computing resources between different customers; Pay per use: customers pay for the service instead of buying software licenses and hardware; On-demand usage/flexibility: cloud services can be used almost instantly and can easily be scaled up and down; External data storage: customers' data is stored externally at the location of the cloud computing provider; Highly reliable services: clouds manage themselves in case of failures or the performance degradation [15] .

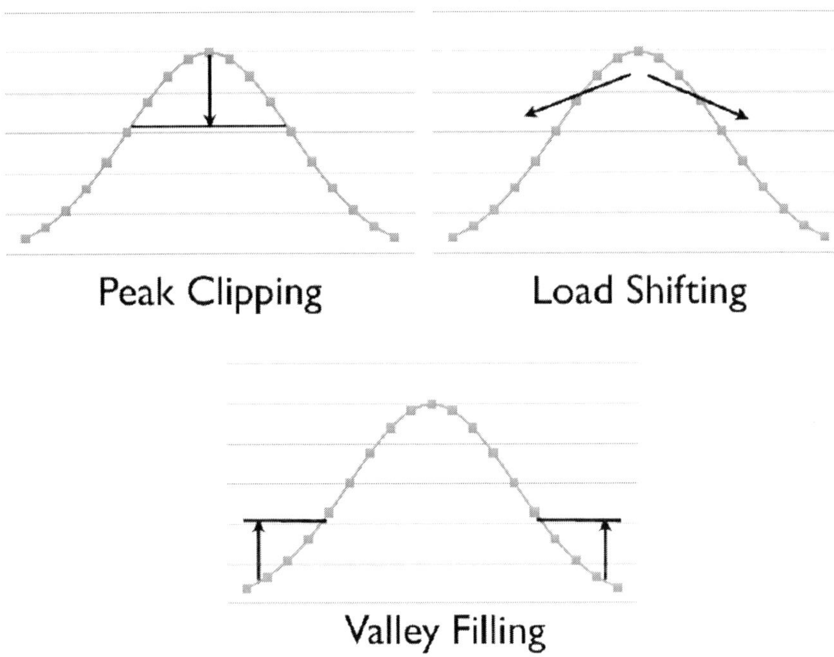

Figure 5. Strategies for load management [12] .

One of the most common applications for real-time data in manufacturing and process industries is SCADA, supervising remote processes over a network. With the growing popularity of cloud computing, many engineers and managers in the automation sector are looking at the possibility of using the cloud for SCADA. SCADA systems have evolved over time and have followed the progress of computing in general. As many view cloud computing as the next logical step in this evolution, enthusiastic visionaries foresee a fourth, "cloud" generation of SCADA, where an entire control system would be running in the cloud [16] . The cloud is creating a revolution in SCADA system architecture because it provides very high redundancy, virtually unlimited data storage, and worldwide data access all at a very low cost [17] .

The cloud computing can support SCADA applications in two ways: The SCADA application is running on-site, directly connected to the control network and delivering information to the cloud where it can be stored and disseminated. The control functions of the SCADA application are entirely isolated to the control network. However, the SCADA application is connected to a service in the cloud that provides visualization, reporting, and access to remote users. These applications are commonly implemented using public cloud infrastructures (PaaS cloud service). The SCADA application is running entirely in the cloud and remotely connected to the control network. The controllers are connected via WAN links to the SCADA application running entirely in the cloud. These applications are commonly implemented using private or hybrid cloud architectures (IaaS cloud service) [18] [19] .

Because of wide ranging variability of the entities in Smart Grids, there is a very high level of potential complexity in finding the optimal solution for each different Smart Grid. Smart Grid will eventually be deployed across all types of infrastructure using widespread Internet of Services, connecting all smart objects worldwide. It will become the major application domain of the Internet of Things, perhaps even referred to as the Internet of Energy [19] .

6. BARRIERS OF BRAZILIAN ELECTRICAL GRID

Energy management systems in residential, commercial, and industrial sectors can play an important role in reducing peak demand of electric network. Eventually, it can help in reducing overhead and stress on transmission and distribution lines. In many countries, there are various demand response programs implemented for the industrial and commercial sector. With the installation of energy management systems, it becomes possible to perform load control, mainly through the models of RTP (Real Time Pricing) and TOU (Time-of-Use) [20] .

Few demand response programs are used for management energy in the residential sector in Brazil. Direct restriction of loads is the most popular method used to reduce peak demand. But by direct control of the load, the consumer comfort can be compromised. In contrast, by the method of load displacement, the loads that have less impact to the consumer lifestyle are displaced outside of peak hours, ensuring a better stability of the network.

To analyze and consider the use of the concept of demand response is important to understand the load models "Hung" on the grid, for such function Smart Grid technologies play a fundamental role. This identification in conjunction with a proper communication between the consumer and the energy utilities and demand management of domestic charges are indispensable factors for high efficiency power management systems.

7. FINAL CONSIDERATIONS

Energy management systems, when developed in a context Smart Grid, have their functions enhanced, in regards of the technologies making up the system. Among the main features of an energy management system, the following stand out as the most enhanced:

- Real-time performance monitoring;
- Information to compose predictive maintenance planning;
- Energy management;
- The efficiency of the power system;
- Financial economics, avoiding any sanction from the energy utilities;
- To keep the tolerance of voltage and current to the extent allowed;
- Continuous monitoring for power quality;
- Load control and management with appropriate methodology;
- To keep consumption with the limit signed;
- Database of information that can assist future decision-making;
- Identify and correct the causes of energy disorder to avoid recurrences.

Among the main Smart Grid technologies, the most important and that directly impacts the design of an energy management system are the solutions on cloud computing. Cloud computing has established itself as an adequate means to provide resources to customers, primarily in energy management systems, with access to a large amount of information and computer storage. With cloud computing, customers do not have to manage and maintain their own information technology (IT), and are not bound to its local resources which often are limited. However, for customers and energy utilities, making sure that your cloud services are usable, an appropriate level of guarantees of Quality of Service (QoS) is needed.

In recent years, the creation of solutions in Data Center Networks (DCNS) came with rapid growth in scale and complexity, making possible hosting large applications, known as cloud hosting. This growth imposes enormous challenges to update the current datacenter infrastructure, especially considering a scenario of Smart Grid with cloud computing solutions, broadly used in energy management systems.

The proliferation of the adoption of cloud computing solutions in recent years is driven by the potential for obtaining benefits such as reduced costs, greater agility, and better use of resources. However, there are many challenges to ensure the success of these cloud-based services, and these need to be understood and managed before the major use in concepts such as Smart Grid.

However, the major current infrastructures are owned by a large number of Internet Service Providers (ISPs); and it is difficult to adopt new architectures without the agreement of all parties concerned. This includes the standardization of communication protocols and creating regulations for wide use of cloud computing solutions.

ACKNOWLEDGEMENTS

The authors would like to thank the Coordenação de Aperfeiçoamento de Pessoal de Nível Superior (CAPES), the Concelho Naional de Desenvolvimento Científico e Tecnológico (CNPq), the Fundação de Amparo à Pesquisa do Estado de São Paulo (FAPESP), the Departamento de Comunicações (DECOM), the Faculdade de Engenharia Elétrica e de Computação (FEEC), and the Universidade Estadual de Campinas (UNICAMP), for their support in the development of this research.

REFERENCES

1. Eissa, M. (2012) Energy Efficiency—The Innovative Ways for Smart Energy, the Future towards Modern Utilities.
2. Ireshika, M.A.S.T. (2014) Home Energy Management System. Universitetet I Agder, Kristiansand & Grimstad.
3. Khodaei, A., Shahidehpour, M. and Bahramirad, S. (2011) SCUC with Hourly Demand Response Considering Inter-temporal Load Characteristics. IEEE Transactions on Smart Grid, 2, 564-571.
4. Du, P. and Lu, N. (2011) Appliance Commitment for Household Load Scheduling. IEEE Transactions on Smart Grid, 2, 411-419.
5. Gatsis, N. and Giannakis, G.B. (2012) Residential Load Control: Distributed Scheduling and Convergence with Lost AMI Messages. IEEE Transactions on Smart Grid, 3, 770-786.
6. Li, J., Chung, J.Y., Xiao, J., Hong, J.W.-K. and Boutaba, R. (2011) On the Design and Implementation of a Home Energy Management System. Proceedings of the 6th International Symposium on Wireless and Pervasive Computing, Hong Kong, 23-25 February 2011, 1-6.

7. Stephens, J., Wilson, E.J. and Peterson, T.R. (2015) Smart Grid (R) Evolution. Cambridge University Press, Cambri-dge.

8. Jiang, T., Yu, L. and Cao, Y. (2015) Energy Management of Internet Data Centers in Smart Grid.

9. Choi, C.-S., Ian, J.I., Park, W.-K., Jeong, Y.-K. and Lee, I.-W. (2011) Proactive Energy Management System Archi-tecture Interworing with Smart Grid. Proceedings of the IEEE 15th International Symposium on Consumer Electronics, Singapore, 14-17 June 2011, 1-4.

10. Park, K., Kim, Y., Kim, S., Kim, K., Lee, W. and Park, H. (2011) Building Energy Management System based on Smart Grid. Proceedings of the IEEE 33rd International Telecommunications Energy Conference, Amsterdam, 9-13 October 2011, 1-4.

11. Balijepalli, V.S.K.M., Pradhan, V., Khaparde, S.A. and Shereef, R.M. (2011) Review of Demand Response under Smart Grid Paradigm. Proceedings of the 2011 IEEE PES International Conference on Innovative Smart Grid Techno-logies-India, Kollam, 1-3 December 2011, 236-243.

12. Paracha, Z.J. and Doulai, P. (1998) Load Management: Techniques and Methods in Electric Power System. Proceedings of the International Conference on Energy Management and Power Delivery, Singapore, 3-5 March 1998, 213-217.

13. Medina, J., Muller, N. and Roytelman, I. (2010) Demand Response and Distribution Grid Operations: Opportunities and Challenges. IEEE Transactions on Smart Grid, 1, 193-198.

14. Bashir, A.K., Ohsita, Y. and Murata, M. (2015) Abstraction Layer Based Distributed Architecture for Virtualized Data Centers. Proceedings of the Sixth International Conference on Cloud Computing, GRIDs, and Virtualization, Nice, 22-27 March 2015, 62-67.

15. Frey, S., Disch, S., Reich, C., Knahl, M. and Clarke, N. (2015) Cloud Storage Prediction with Neural Networks. Proceedings of the Sixth International Conference on Cloud Computing, GRIDs, and Virtualization, Nice, 22-27 March 2015, 68-72.

16. Mcilvride, B. (2012) Will SCADA Envolve to the Cloud? http://real-timecloud.com/

17. Combs, L. (2011) Cloud Computing for SCADA. http:// www . controleng.com/single-ar

18. Conway, G., Carcary, M. and Doherty, E. (2015) A Conceptual Framework to Implement and Manage a Cloud Computing Environment. Proceedings of the Sixth International Conference on Cloud Computing, GRIDs, and Vir-tualization, Nice, 22-27 March 2015, 138-142.

19. Markovic, D.S., Zivkovic, D., Branovic, I., Popovic, R. and Cvetkovic, D. (2013) Smart Power Grid and Cloud Com-puting. Renewable & Sustainable Energy Reviews, 24, 566-577.

20. Albadi, M.H. and El-Saadany, E.F. (2007) Demand Response in Electricity Markets: An Overview. Proceedings of the 2007 IEEE Power Engineering Society General Meeting, Tampa, 24-28 June 2007, 1-5.

2

CHAPTER

ENERGY EFFICIENCY OF ELECTRIC VEHICLES – ENERGY SAVING AND OPTIMAL CONTROL TECHNOLOGIES

Guoqing Xu[1, 2], Chunhua Zheng[1, 2], Yanhui Zhang[1], Kun Xu[3] and Jianing Liang[1]

[1] Automotive Electronics Research Center, Shenzhen Institutes of Advanced Technology, Chinese Academy of Sciences, Shenzhen, China

[2] Department of Mechanical and Automation Engineering, The Chinese University of Hong Kong, Hong Kong, China

[3] Shenzhen College of Advanced Technology, University of Chinese Academy of Sciences, Shenzhen, China

1. INTRODUCTION

Electric vehicles (EVs), including pure electric vehicles (PEVs or EVs), hybrid electric vehicles (HEVs), and fuel cell vehicles (FCVs), have attracted worldwide attention increasingly from governments, industry, and public. Though EVs have shown advantages over internal combustion engine (ICE) vehicles in terms of environmental friendliness and energy efficiency, there are still many technical challenges impeding the market share growth.

In this chapter, the energy-related issues for EVs will be discussed in detail. The primary aim of this chapter is to give a clear and systematic understanding of the energy processing and control for EVs and to introduce some latest energy saving and optimal control technologies and present the key results regarding this area. The overall technical architecture and the overview of the state of the art will be introduced. Then some advanced and newest technologies of energy savings and optimal control for EVs, such as the high-efficiency traction motor, intelligent battery management system, regenerative braking with high energy recovery, and the vehicle energy management, will be included and discussed as well as the key results. The detailed contents of this chapter are as follows:

2. INTRODUCTION OF EVS

2.1. EVs

The transportation sector is the largest consumer of oil and it has grown at a higher rate than any other sector in recent decades [1]. Increases in transportation have caused pollution emissions because of the utilization of ICEs, and that have consequently caused some environmental problems, which must be prevented to maintain the quality of life. Thus, many governments have developed more stringent standards for the pollution emissions of vehicles. In order to achieve better fuel economy and lower emissions, both academia and automotive industry have conducted research on improving the efficiency of powertrains, developing other energy sources, and changing the concept of the conventional powertrains to EVs including PEVs, HEVs, and FCVs. Among them, PEVs use electric motors for traction and batteries or supercapacitors for energy storage systems. FCVs take the fuel cells as the energy sources, however, the fuel cells do not have the energy recovery function which the batteries and the supercapacitors have. PEVs and FCVs have many outstanding advantages over ICE vehicles such as zero emission, independence from petroleum, higher efficiency, and quiet operation [2]. As a power source, a fuel cell system (FCS) has a relatively slow power response and the ICE or the fuel cell cannot recover the vehicle braking energy. Thus, the size of the FCS or the ICE will be increased if the FCS or the ICE is the only power source in a vehicle. The secondary power source which has a relatively quick power response and can recuperate the braking energy is needed, and a battery or a supercapacitor could be one of the candidates for the secondary power source. Hybrid powertrain structure is the solution of the above requirements. A fuel cell hybrid vehicle (FCHV) or an HEV can provide sufficient power during its acceleration and can recuperate the kinetic or potential energy of the vehicle during braking by the hybridization.

2.2. The state of the art of energy saving and optimal control technologies for EVs and the technical challenges

For the traction motors, the optimization of different machines is still the most important means to pursue higher power density and higher efficiency. In order to reduce the overall weight, volume, and cost, the integrated motor drive becomes more and more popular which integrates the circuit of the charger into other power electronic circuits existing in EVs. In addition to the three-phase AC motors, switched reluctance motor (SRM) has attracted increasing attention recently due to its low cost, simple structure, good controllability, high efficiency, and wide operation range. SRMs have been extensively applied to the electric drive systems of EVs.

For the BMS of EVs, the main and key difficulties are the voltage balance and the battery SOC estimation. The voltage balance is dependent on the accurate estimation of the battery SOC. Thus, the battery SOC estimation is not only for the voltage balance but also for achieving a good energy management result. Currently, the ampere-hour (Ah) model is one of the most widely used methods in this area.

For the RBS, in spite of the significant benefits of RBS, there are still technical challenges. The battery conditions, including the SOC and temperature, affect the energy recovery. When the SOC is too high, the regenerative braking energy may not be fed back to the battery. The hydraulic braking system functions instead of the RBS in such case. Additionally, the hydraulic braking system is reserved in case of failure of the electric braking actuator. Therefore, the hybrid brake system (HBS) that consists of both the hydraulic and the regenerative braking systems is the main architecture in this area, which is further divided into the parallel type and the series type. The HBS needs an appropriate coordination mechanism between the two subsystems to achieve multi-objectives, i.e., safety, energy efficiency, and brake comfort. Various control methodologies including the rule-based, fuzzy logic-based, robust control-based, and neural network-based approaches have been proposed and investigated in the literature. Nevertheless, the HBS needs further studies due to the model nonlinearity, parameter uncertainty, external disturbances, and the interaction with the driver's operation on the brake pedal. From the point of view of the safety and energy savings, emphasizing the electric braking force is a significant trend in future.

For the energy management of EVs, rule-based energy management strategies are usually used currently because of the simplicity. Other advanced strategies, for example optimal control theory-based energy management strategies, are still on the research step due to several unsolved problems such as lack of the memory for big data saving and low speed of the processor. In addition, the research on the energy management strategies using the prediction of future driving condition is active currently, as the driving route of the vehicle is not fixed in the real-world driving.

3. ELECTRIC DRIVING TECHNOLOGIES

3.1. Introduction of EV traction motors

The traction motors, working as the heart of powertrains, are the most important key components of modern EVs. Unlike conventional motors, traction motors and drive systems for EVs are under special demands, such as compact structure, light weight, high reliability, high efficiency, low noise, low vibration, wide operating range, and most importantly, low cost. With the harsh requirements, innovative concepts are used to design the EV traction motors, which include the innovations on conventional motor structure, principle, design, performance, and manufacturing technology. In order to achieve high power/torque densities, high reliability, and wide operating range, a lot of new technologies such as water-cooled frame structure, internal oil-cooled stator structure, thin-shell frame, and ultra-short and solid filled-in end-turns will be utilized. Due to the natural disadvantages, such as complicated brush assembly, poor reliability, and the need of maintenance, conventional brushed DC machines are not suitable for larger scale commercial application in vehicle business. Modern vehicles normally use AC machines without brushes, namely, induction machines, switched reluctance machines, and permanent magnet AC machines.

To pursue a high power density of machines, the major task is to increase the peak output torque, rather than the rotating speed, because the speed may be limited by applications and manufacturing technology of the powertrain. The size and weight of machines are mainly related to the peak output torque, not peak output power. The peak torque depends on the saturating level of magnetic path and the current in the winding. Designers have to endlessly optimize the structure of magnetic path to achieve better usage of ferromagnetic materials and search for peak torque with smaller winding current. For the similar heat dissipation condition, the smaller winding current results in less conductor materials and the ferromagnetic materials surrounding them and consequently leads to less resistance losses in the conductor and less iron losses in the ferromagnetic materials. Hence, the higher power density requires the higher efficiency, while the higher efficiency normally results in the higher power density. Under certain situations, the higher power density can also be achieved with higher current density, which requires promotions on the heat resistance of materials and improvements of heat dissipations. However, such solutions demand for better materials and manufacturing technologies. The total power loss will be the largest in induction machines, smaller in reluctance machines, and the smallest in permanent magnet (PM) machines. The efficiency of the machines will follow the reverse sequence. Thus, the PM machines come with the most potential for higher power density. In fact, the optimization of different machines is still the most important means to pursue higher power density and higher efficiency.

3.2. Electric drive based on integrated technology

Transport electrification has been considered as one of the most promising solutions to the urgent challenges facing our human beings, such as global warming, environmental pollution, energy crisis, and so forth [3]. Most recently, with the advent of the vehicle-to-grid (V2G) technology [4], the gridable EVs that can be directly connected to the utility grid have been expected to function as distributed energy storage system during the parking period, so as to balance the demand and supply of the electricity market in the future. Generally, gridable EVs include battery EVs, plug-in hybrid EVs [5, 6], and range-extended EVs [7].

In contrary to the traditional EVs, the gridable EVs are equipped with plugs and on-board chargers. By connecting the plug to the utility grid, the electrical energy can flow into the on-board batteries through the on-board charger. Moreover, the energy deposited in EVs can also be fed back to power grid when the bi-directional on-board charger is engaged.

In gridable EVs, the high power energy conversion systems are installed, for example, motor drive system, low voltage DC/DC converter, electrical air conditioning system, battery charger system, and grid-connect inverter shown in Figure 1. Many complex power electronics and electrical systems bring the serious challenges from system layout, installation, weight, cooling, electromagnetic compatibility, and cost, etc. The diverse needs for the electromechanical systems of EVs can be satisfied by the development of integrated technology.

In order to reduce the overall weight, volume, and cost, integrated motor drive becomes more and more popular to integrate the circuit of the charger into other power electronic circuits existing in EVs. In the previous publications, the circuits of motor drive and the motor are involved to construct integrated on-board chargers [8, 9]. Due to the neutral point of AC motors, one terminal of AC line can be conveniently connected to the neutral point, and the three-phase windings can function as the couple inductances in the boost converter circuit. However, the boost converter topology requires the input voltage lower than the output voltage. Thus, such kind of integrated charger is more feasible in countries and regions where 110 V AC power is adopted. In order to extend the range of the input voltage, the buck converter has been added to the battery side.

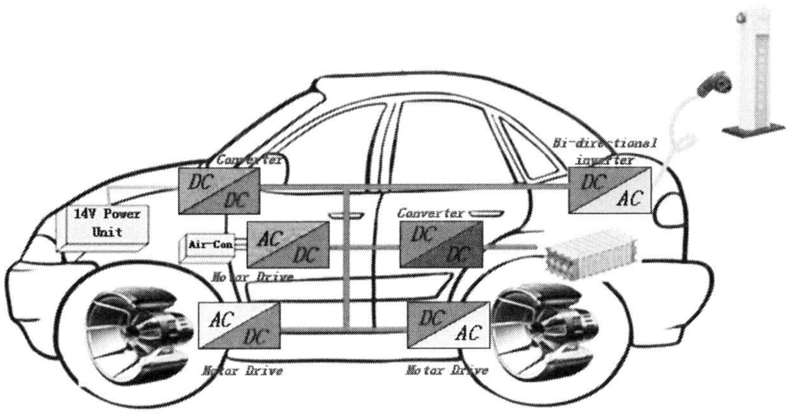

Figure 1. Main electrical energy conversion system in EVs

In addition to the three-phase AC motor, SRM has attracted increasing attention recently due to its low cost, simple structure, good controllability, high efficiency, and wide operation range [10, 11]. SRMs have been extensively applied to the electric drive systems of EVs. The purpose of this study is to propose a novel integrated SRM drive with the bi-directional inverter to reduce the cost, volume, and weight for EV applications.

3.3. Integrated technology of switched reluctance motor drive

3.3.1. Conventional SRM and its converter

SRMs, which are non-rare earth permanent magnet motor, have been recognized as the considerable candidates for EV and HEV applications. It has salient poles on both rotor and stator, but only the stator carries windings. The torque is produced by the alignment tendency of poles. The rotor will shift to a position where the inductance of the excited winding is maximized. Therefore, the advantages of SRMs can be summarized as simple construction, low manufacturing cost, and outstanding torque characteristics. The structure of SRMs is shown in Figure 2.

When the rotor position enters into inductance increasing region, this phase winding is injected with the phase current. SR motor operates at motoring mode and produces positive torque. Contrarily, SR motor operates at regeneration mode and produces negative torque. Motoring and regeneration modes of SRM are shown in Figure 3.

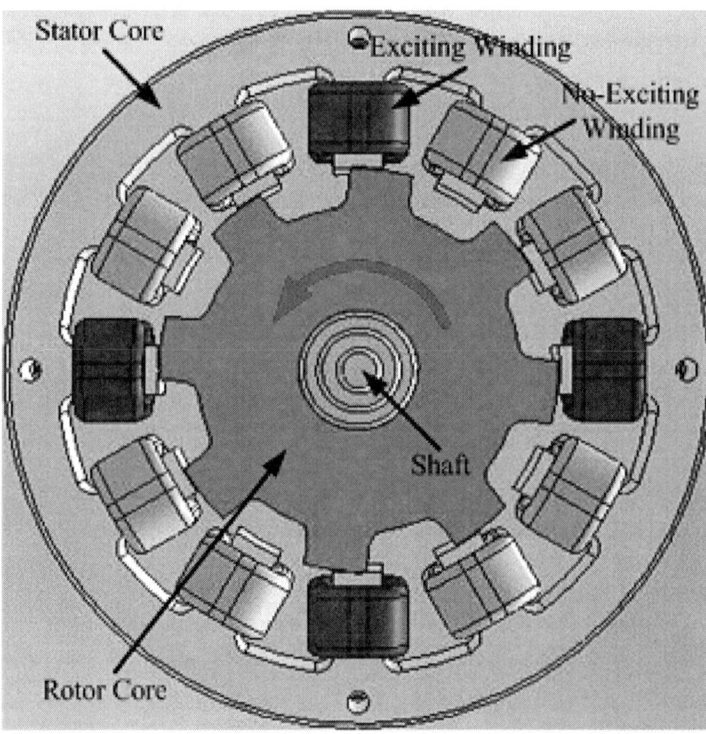

Figure 2. Excited phase winding in 12/8 3-phase SRM

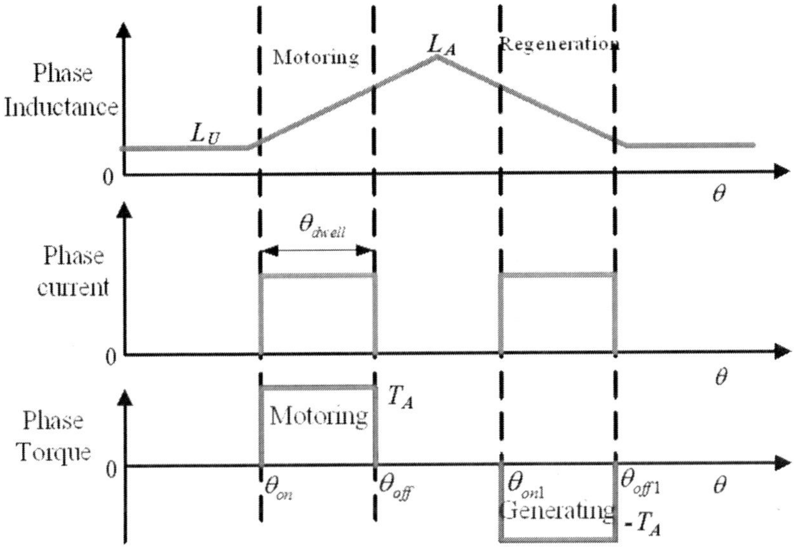

Figure 3. Motoring and regeneration modes of SRM

For EV applications, the conventional SRM drive uses a battery source and a large capacitor in the front-end as shown in Figure 4. This capacitor helps the battery to keep a steady DC-link voltage, acting as a low-pass filter. Another function of DC-link capacitor is to store the magnetic field energy when the SR motor is demagnetized.

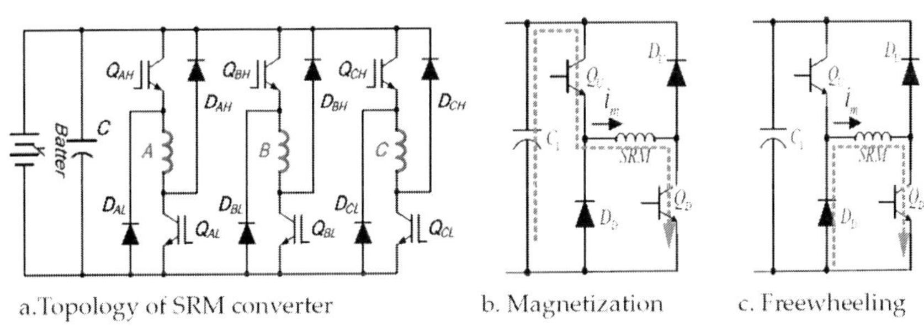

Figure 4. Conventional 3-phase SRM drive and its operation modes

The asymmetric bridge converter consists of two power switches and two diodes per phase, which can support independent control of each phase and handle phase overlap. The asymmetric converter has three modes, which are defined as magnetization mode, freewheeling mode, and demagnetization mode as shown in Figure 4 (b-d).

3.3.2. Integrated switched reluctance motor converter

In the conventional EV application, the on-board battery charge and the grid-connect inverter are independent devices. To reduce the size, weight, and cost, the integrated SRM drive with the bi-directional inverter is introduced [12]. The integrated converter can be operated as: SRM drive, battery charger, and grid-connect inverter. In order to improve the coefficient of utilization, the power switches, inductors, capacitors, and controller are multiplexed in the same circuit. For city vehicle application, the operation of driving mode, battery charging mode, and grid-connect mode can be divided by Time-Division Multiplexing (TDM) method. The integrated drive system can be served the activity plan of daily as shown in Figure5. The integrated system can be seamless transferred to each application. Of course, more complicated case can be supported by this method.

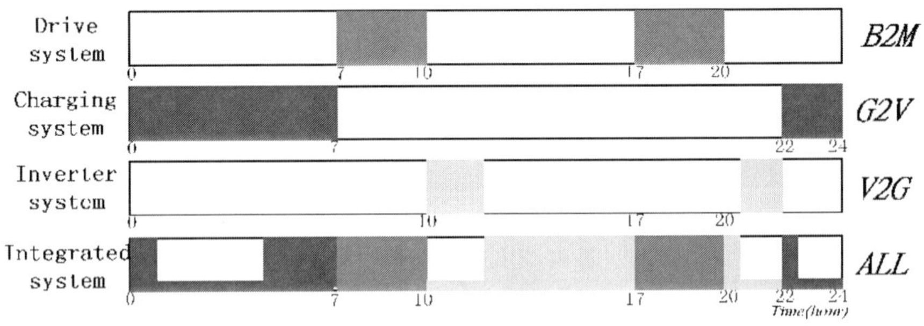

Figure 5. The activity plan of different application in working day

The battery charger is a device used to put grid energy into the battery by forcing an electric current through it. It consists of the rectifier, the boost converter, and the buck converter. The grid connect inverter is a device used to put battery energy into the grid. It is made up of the boost converter and inverter bridge. Due to the power factor requirement, the power factor correction is required to perform. The topology of three circuits is shown in Figure 6. In order to build the integrated motor drive, all the functions of the three devices should be included.

A novel integrated SRM drive with the bi-directional inverter is introduced in Figure 7 [12]. Compared with the conventional asymmetric converter, the two lines from the plug outlet are connected to the cathode of the lower diode in phase A and B. A relay is added between the switch Q_{CH1} and the switch Q_{CH2} to divide the voltage stage of DC-link and battery. In order to reconstruct the bi-directional converter, two free-wheeling diodes of phase C's are replaced by two power switches with parallel protection diodes.

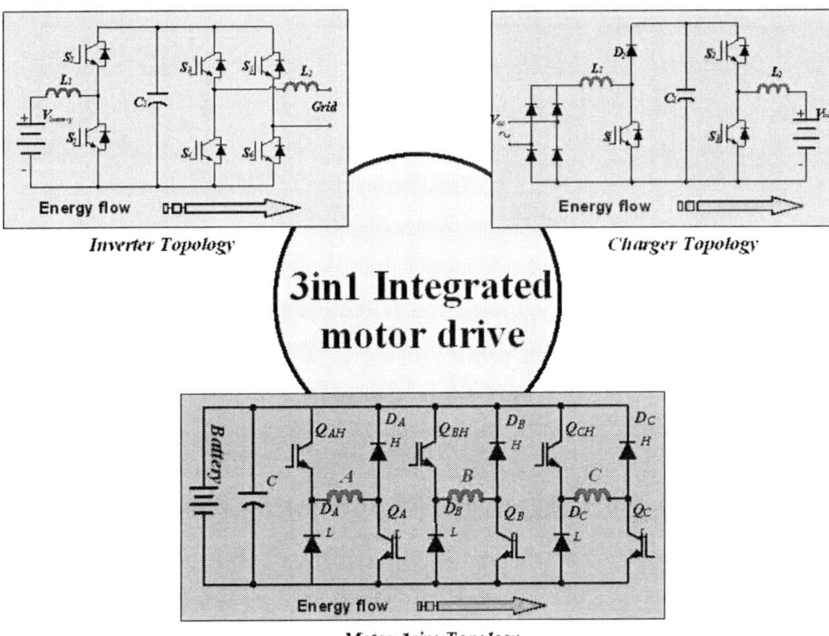

Figure 6. The circuit topology of motor drive, charger, and inverter

Obviously, two stages of the bi-directional converter are suitable for the battery charger application due to the wide range of the battery voltage. The bulk capacitor of the motor drive is designed to separate two parts and serve two independent voltage stages. In the driving mode, the two capacitors are paralleled to increase the capacitance value. In the bi-directional converter mode, the two capacitors are divided: one is for the boost voltage stage, and the other is for the battery voltage stage. In the battery charger mode, the relay is opened and the proposed integrated SRM drive is constructed by the bridgeless power factor correction (PFC) front-end and the buck-boost converter. The bridgeless PFC front-end can improve the efficiency of AC-DC conversion. The key point of this integrated SRM drive is that it applies the motor winding as the boost inductance and the buck inductance [12]. In the grid-connected inverter mode, the relay is similarly opened. The bridgeless PFC frond-end is operated as a single phase inverter, and the full bridge converter of the phase C is run as a boost converter in the opposite direction [12].

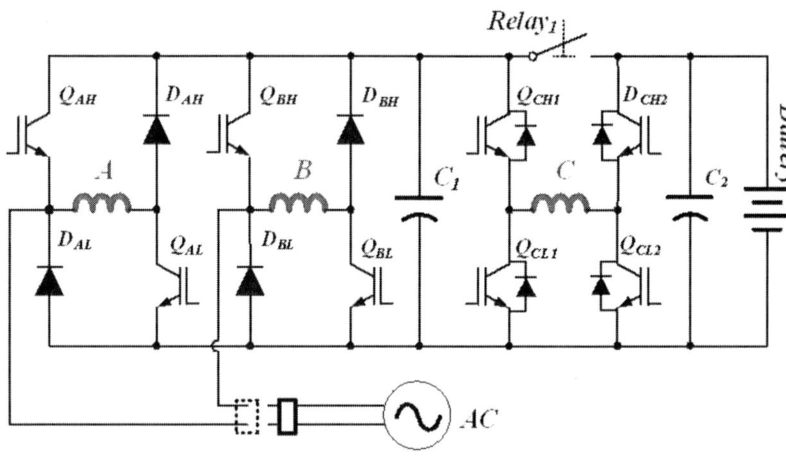

Figure 7. A novel integrated SRM drive with the bi-directional converter

3.3.3. Control strategy of the integrated motor drive

For the normal inverter, the inductance is only like a static transformer. However, in the novel integrated drive system, the inductance of the integrated inverter is related to the phase winding of SRM, which varies according to the rotor position and phase current. In this application, the phase inductance is limited by the primary characteristic of the SRM drive. Thus, three problems can be found: the first is that the inductance of the SRM is changed by the rotor position; the second is that the relative space angle between phase windings is 120 electrical degrees; and the last is that the phase inductance has effects of magnetic saturation.

In the motor driving mode, the rotor performs the rotational motion according to the instructions. However, unlike the driving mode, the rotor position of the battery charging mode and the grid-connect mode prefers not to rotate. The rotation and vibration only increase the loss of energy conversion. Two questions appear: how to fix the rotor at still and how to find the suitable position.

For the fixed position, two methods can be selected: the mechanical brake method and the zero torque control method [11]. For the mechanical method, the rotor will be locked by the mechanical structure. The motor torque will be counteracted, but the mechanical components increase the cost and space. Due to satisfy the power factor correction and battery charging, the sine wave current is injected in the phase A or B, and phase C is flown in the direct current. Therefore, the zero torque control method cannot be applied in this converter topology [13]. Thus, the mechanical method is used, and the advantage is that the rotor position is easy to set to any position.

For finding the suitable rotor position, the aligned position of the phase A is defined as zero. Ignoring the influence of mutual inductance, the inductance of the SRM is a periodic function of the rotor position, and it can be simplified and expressed using Fourier series as follows:

$$L(\theta, i) = \sum_{n=0}^{\infty} L_n(i)\cos(n\theta) \qquad (n = 0, 1, 2...) \tag{1}$$

where *Ln* (i) is the amplitude of the Fourier components, whereas θ_A, θ_B, and θ_C are electrical degrees of each phase. In a three phase SRM, the axes of the three-phase windings are spaced 120° electrical degrees from one another, that is

$$\theta_A = \theta \tag{2}$$

$$\theta_C = \theta + {2\pi}/{3} \tag{3}$$

$$\theta_B = \theta + {4\pi}/{3} \tag{4}$$

For the bridgeless PFC front-end circuit, a boost inductor is needed to boost-up the DC-link voltage. However, in the novel integrated drive system, the winding of phase A or B is alternated to replace the boost inductor in the different half-cycle of sinusoidal voltage. Obviously, the inductor of the phase A and B should be equal in value to make the circuit symmetrically. Therefore, the charging position of the novel integrated system must meet the inductor requirement.

$$L(\theta_A, i) = L(\theta_B, i) \tag{5}$$

The inductance of the phase A and B can be calculated from (1), (4), and (5):

$$\theta_A = -\frac{2\pi}{3} + k\pi, \qquad (k = 0,1,2..) \tag{6}$$

So, meet the conditions in case of rotor position is

$$\theta_A = \frac{\pi}{3}, or \quad \theta_A = \frac{4\pi}{3}. \tag{7}$$

In both cases, the difference between the inductance of the phase A and B is clearly indicated in Figure 8. The inductance of the phase A and B at π/3 is higher than that at 4π/3. However, the inductance of the phase C reaches the maximum value at 4π/3 and the minimum value at π/3. The selection of the charging position should be decided by the real application. Due to the phase C winding used by DC/DC converter, the inductance of the phase C will affect the dynamic response ability.

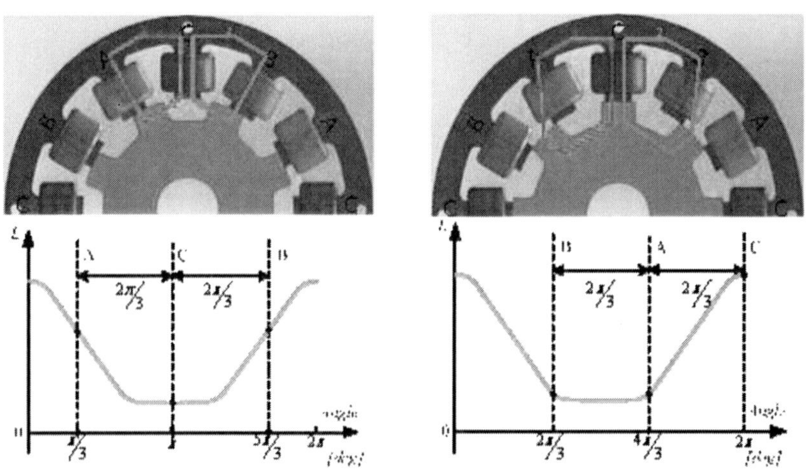

(a) $\theta_A = \pi/3$ (b) $\theta_A = 4\pi/3$

Figure 8. Rotor position of phase A in the bi-directional inverter mode

The block diagram of the novel integrated SRM drive system is shown in Figure 9. In the driving mode, the battery supplies the stored energy to drive SR motor, and the power flow is through the battery, integrated converter, and SR motor. When the outlet is connected to the grid, the vehicle management system should decide operation mode of the integrated drive system. If charging mode is executed, the asymmetric converter of the phase A and B is operated as a bridgeless PFC converter to meet the requirement of the PFC and DC-link voltage. And the full bridge converter of the phase C is implemented as a buck converter to manage the battery charging strategy. The power flow is followed the grid, SR motor, proposed converter, and the battery.

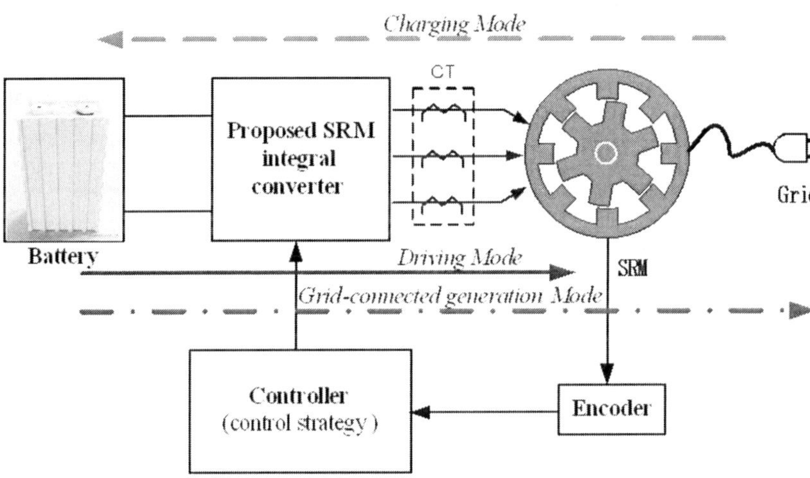

Figure 9. Block diagram of the integrated system

Due to the topology of the bridgeless front-end, the positive half cycle of the line current is injected into the winding of the phase A, and the negative half cycle of the line current is injected into the winding of the phase B. However, the input line current waveform is also sinusoidal. At the rotor position of $4\pi/3$, the inductance of the phase C reaches the maximum value and the torque of the phase C is zero. Because the phase A and B are conducted in turn, the total torque of the rotor is not equal to zero in Figure 10. The power factor of 220Vac input is 0.99.

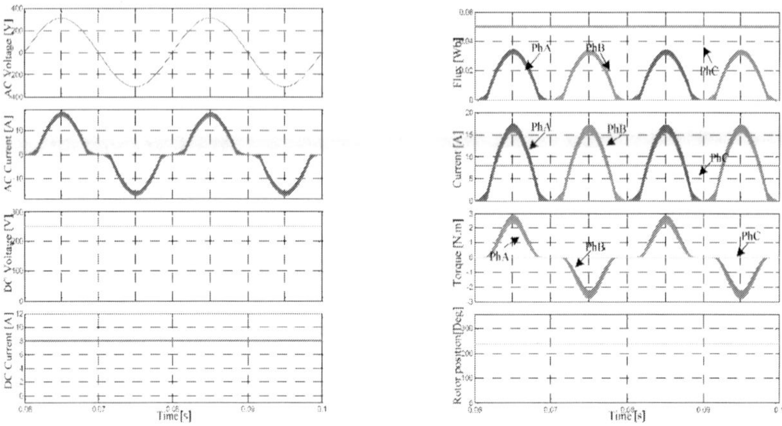

(a) Voltage and current (220[Vac]) (b) Current, flux and torque (220[Vac])

Figure 10. Simulation results of novel integrated system in charging function (220V)

If the inverter mode is carried out, the quadrant DC/DC converter of the phase C is implemented as a boost converter which meets the DC-link voltage requirement. The asymmetric converters of the phase A and B are acted as the inverter bridges. The power flow passes the battery, converter, SR motor, and the grid.

In the inverter function (Figure 11), the current of the phase C is the reverse direction of charging function's. Thus, the phase current and flux of the phase C are negative. Since the current of the phase A and B is unidirectional, the flux of them is positive. For the battery side, the battery discharges the energy to the DC-link. The current and voltage of power grid have 180 degrees.

4. ADVANCED BATTERY MANAGEMENT TECHNOLOGIES

4.1. Introduction of battery management system (BMS)

The lithium-ion battery has attracted special attention for HEVs and EVs owing to its advantages of high specific power, high energy density, no memory effect, and long durability. What is more, research is yielding a series of improvements to traditional manufacturing technology focusing on power density and intrinsic safety of battery.

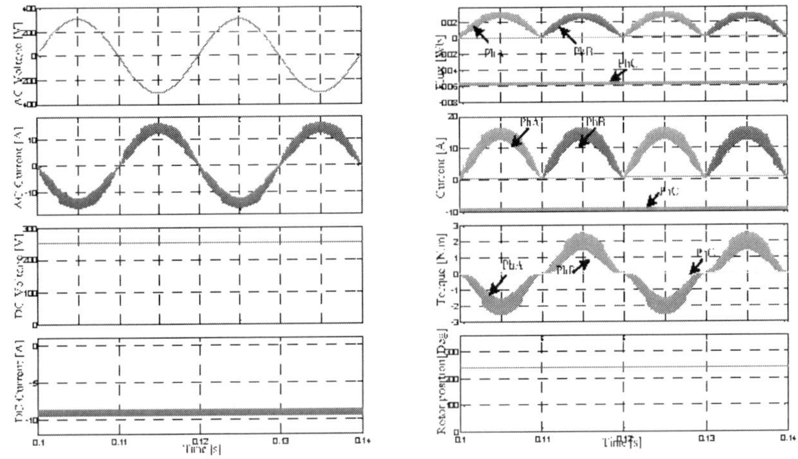

(a) Voltage and current(220V) (b) Current, flux and torque(220V)

Figure 11.Simulation result of proposed integrated system in inverter function (220V)

A BMS is an important means for raising the intrinsic safety and durability of batteries. The main parameters of the batteries are monitored with the BMS including the voltage, current, temperature, and also the leakage detection and SOC & state of health (SOH) condition. The maximum mileage and the charging control algorithm can be achieved based on the battery voltage, current, and temperature under the maximum output power control algorithm. The real-time communication is achieved between the CAN bus interface and the vehicle master controller, motor controller, energy control system, and automotive display systems. The BMS is a bridge between the battery and drivers, and it plays a critical role for the EV performance. The diagram of the BMS is shown as Figure 12.

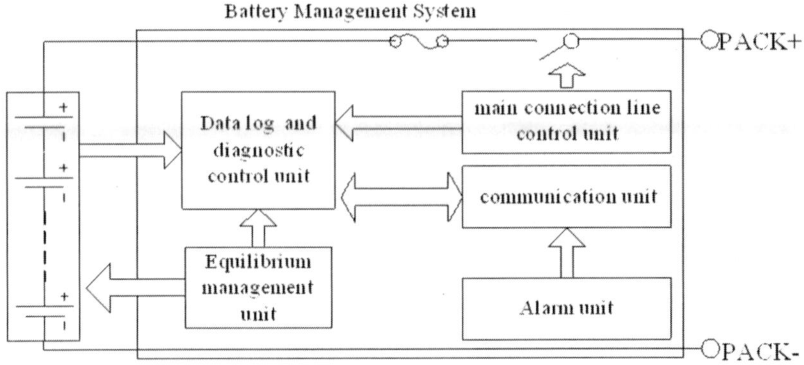

Figure 12. Diagram of the BMS

4.2. SOC estimation of Lithium-ion batteries

Many ways of SOC estimation have been proposed in the literature [14]. According to the main detection parameters, they are classified into voltage model, resistance model, current model, and intelligent algorithm model [15]. The open circuit voltage (OCV) provides the useful information on the intrinsic characteristics of the battery and it declines proportionately with the energy expenditure [16]. However, it does not apply to frequent repeated conditions of charge/ discharge. Moreover, this similarly situation also occurs in the resistance model. Ah model is one of the most widely used methods, and it is an iterative algorithm of dynamic current with time integral. The main drawback is the initial SOC (SOC0) cannot be estimated. Additionally, the intelligent algorithm model, such as Kalman filter model, Extended Kalman filter model, and Artificial neural network model, is highly dependent on the quantity of battery model and input variables, which greatly influences the accuracy and computational complexity. The hybrid model of OCV model and Ah model is considered to be easy to accomplish. Additionally, it is easily implemented in EVs and it needs uncomplicated hardware.

4.2.1. Definition of terms related to SOC

Accordingly, when a battery is in discharging step, the formula to calculate SOC as the ratio of releasable capacity, $Q_{releasable}$, relative to the rated capacity, Q_{rated}, during the profiles is as follows:

$$SOC = \frac{Q_{releasable}}{Q_{rated}} \times 100\% = (1 - \frac{Q_{consumption}}{Q_{rated}}) \times 100\% \tag{8}$$

If the discharging step is not from the rated energy, the Ah model of SOC estimation could be expressed as (9) and shown as Figure 13. The real time SOC estimation (SOC (t)) is equal to the plus of the initial SOC (SOC0) and the ration of the consumption capacity with the rated capacity in an operating period τ. I is positive for discharging and negative for charging. To enhance the accuracy of SOC estimation, the energy efficiency denoted as η is considered.

$$SOC(t) = SOC_0 + \Delta SOC = SOC_0 + \eta \frac{-\int_{t_0}^{t_0+\tau} Idt}{Q_{rated}} \times 100\% \tag{9}$$

4.2.2. The function of OCV to SOC0

The lithium-ion battery used in the investigation is $LiFePO_4$/graphite battery. A $LiFePO_4$ battery with nominal voltage of 3.65 V and nominal capacity of 20 Ah is selected.

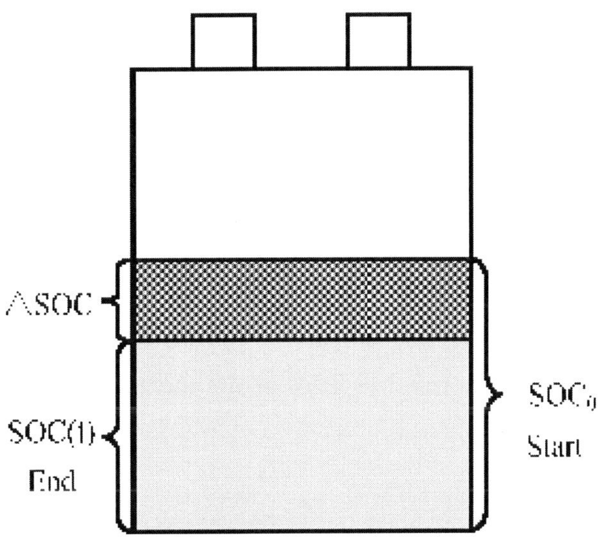

Figure 13. Diagram of relative locations of the initial SOC (SOC0), real time SOC (SOC (t)), and the ration of energy discharged ΔSOC

The OCV is the equilibrium potential difference of an electrode, which depends on the temperature and the amount of active material left in the electrolyte. The OCV can be calculated by Nernst equation [17]. However, the measured voltage at the breaking of the current cannot be denoted as OCV because of the presence of polarization and the relaxation of solid particles. In fact, after the short discharge, the terminal voltage is gradually increased due to relaxation of over potentials and eventually leveled off to an equilibrium potential. Figure 14 illustrates the OCV as a function of SOC after discharging at a room temperature. The verification experiment is conducted to calculate the accuracy of the above estimation method of OCV (see the circles in Figure 14).

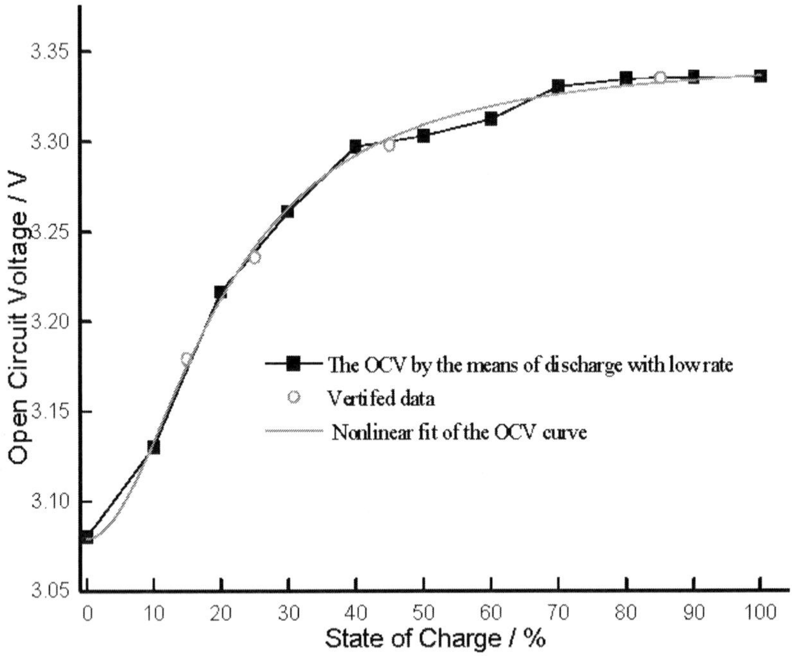

Figure 14. The comparison profiles of the measured voltage and the nonlinear fit

In this study, the quantitative relationship between the SOC and OCV as the logistic function (equation (10)) is proposed. The values of b and p are 20.12155 and 1.97038, respectively, and the correlation coefficient of the fit is 0.997.

$$U = \frac{U(SOC0\%) - U(SOC100\%)}{1 + (SOC_0 / b)^p} + U(SOC100\%) \tag{10}$$

4.2.3. The function of temperature to SOC0

To mitigate the estimation error, the following technique is used in this work. At a certain temperature, the relationship between SOC0 and OCV is obtained and can be established using nonlinear function. dV0 /dT is proportional to the entropy of reaction by the relation dV0 /dT=ΔS/nF, where n is the number of electrons passed in the reaction and F is Faraday's constant, and reflects the influence of the incremental lithium atoms on the ordering of lithium-ion in the host lattice [18].

The battery is discharged to a designed SOC and kept to a temperature of 30°C for 20 hours. This is to impel the degradations of diffusion and migrant to obtain a steady voltage. The temperature experiments focus on the changes of OCV under different ambient temperature conditions of 3°C, 15°C, 25°C, 35°C, and 45°C for 0.5h [19]. The result is shown in Figure 15. The results reveal that the temperature affects the changes of OCV. To reduce the error of temperature upon the SOC0 estimation, the voltage is calculated according to a unified energy unit under a constant temperature, which selects 25°C as a base.

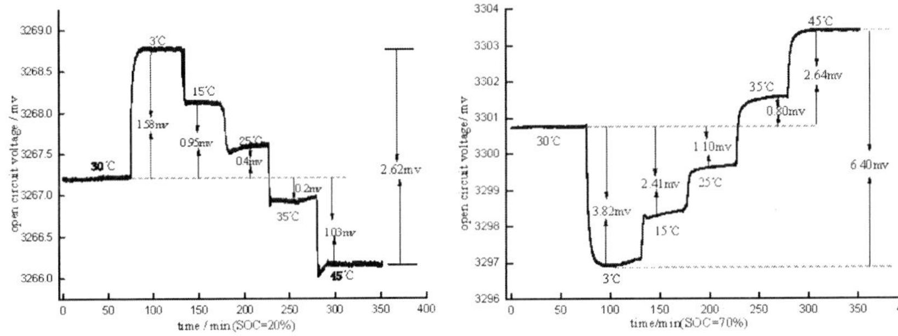

Figure 15. OCV under different temperature (the first and the second subplots respectively correspond to the cases in which SOC=20% and SOC=70%)

Through the above analysis, we can see that the temperature and resting time have significant impacts on the SOC0 estimation. In the present study, the quantitative relationship (equation (11)) exhibits the functions of terminal voltage u, resting time t, and temperature T to SOC0 based on the theory above.

$$SOC_0 = f_3(\Delta t)\exp\frac{1}{f_4(\Delta t)}\left(\frac{f_2(\Delta t)}{u - f_1(\Delta t) - \Delta u(T)} - 1\right) \quad (11)$$

Meanwhile, the temperature correction factor Δu (T), which reflects the influence of the temperature variation on the OCV, is also considered when estimating SOC0.

4.3. Energy Efficiency for intermittent charging-discharging

In order to fully explore the effect of thermal characteristics of lithium-ion battery on the energy efficiency under the continuous process of charging and discharging, the energy balance model under the heat generation behavior is investigated. The energy balance for batteries consists of chemical reactions, ionic mixing, electrical work, and heat transfer with surroundings [20]. Assuming that the temperature distribution in the battery is uniform and the enthalpy of ionic mixing and the radiation between the battery and the environment are neglected, the energy balance function can be written as:

$$q = I^2R + I(\eta + T\frac{\partial V_0}{\partial T}) + MC_p\frac{dT}{dt} \quad (12)$$

$$\eta = V_0 - V \quad (13)$$

Where V0 is the OCV, V is the terminal voltage, q is the heat transfer rate between battery and the surrounding environment, I is the current, R is the resistance, η is the over potential, $IT^{\partial V_0}/_{\partial T}$ is the reversible heat, and Cp is the heat capacity of the battery.

In the charge/discharge cases, we are interested in evaluating how much energy is lost due to the thermal characteristics as a function of internal feature in the battery. If the irreversible reaction energy is ignored and considering the battery back to the initial state after the cycle, the energy balance function can be expressed as follows:

$$e_{charge}+TI_1\frac{\partial V_0}{\partial T}=I_1^2R+\eta I_1+e_{discharge}+TI_2\frac{\partial V_0}{\partial T}+I_2^2R+\eta I_2 \tag{14}$$

To create an energy efficiency model for the charge/discharge cases, we similarly ask what percent of energy loss E_{loss} will be existed in a battery if the battery is charged or discharged at a given current I and a variable power P. The relative definitions of power P, energy loss E_{loss}, and energy stored E_{charge} or discharged $E_{discharge}$ can be visualized in Figure 16.

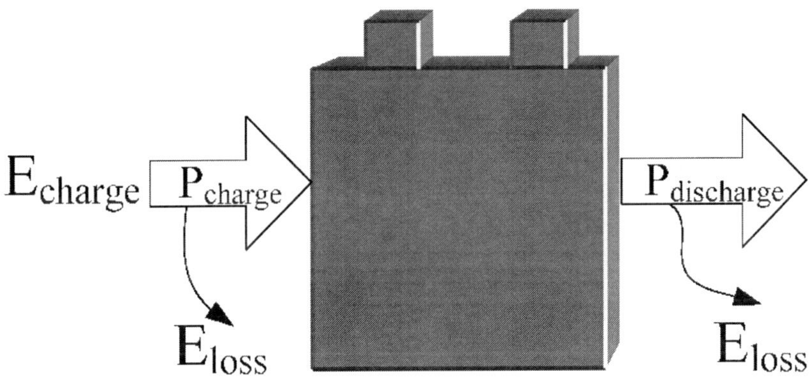

Figure 16. iagram of relative locations in charge/discharge cases

In the process of charging and discharging, the energy lost caused by the battery thermal characteristics can be written as follows:

$$\begin{aligned}\Delta E &= E_{charge}-E_{discharge}\\ &=\int_0^{t_1}(I_1^2R+\eta I_1-TI_1\frac{\partial V_0}{\partial T})dt+\int_0^{t_2}(TI_2\frac{\partial V_0}{\partial T}+I_2^2R+\eta I_2)dt\end{aligned} \tag{15}$$

By using the above function, the energy efficiency of the battery, which is dependent on the thermal feature during constant charge/discharge cycles, is determined by the following:

$$Eff=\left(1-\frac{\Delta E}{\int_0^{t_1}P_{charge}dt+\int_0^{t_2}P_{discharge}dt}\right)\times 100\% \tag{16}$$

The main parameters of the energy efficiency, resistance, and entropy change coefficient can be determined by experiments. The experiments mainly focus on the thermal characteristics of the lithium-ion battery and the analysis of their effect on the energy efficiency under different depth of discharge (DOD) values, as shown in Table 1.

Table 1. Battery test cycle under different DOD values

DOD	SOC	Current (charge)	Current (discharge)
20	40-60	0.3C	0.3C,0.5C,1C
40	30-70	0.3C	0.3C,0.5C,1C
60	20-80	0.3C	0.3C,0.5C,1C
80	10-90	0.3C	0.3C,0.5C,1C

The experimental results are as follows:

i. Resistance

The discharge test at constant current is performed in the experimental setup and a few representatives are selected at 25°C as shown in Figure 17. The results of the resistance obtained for the SOC values are shown in Figure 17.

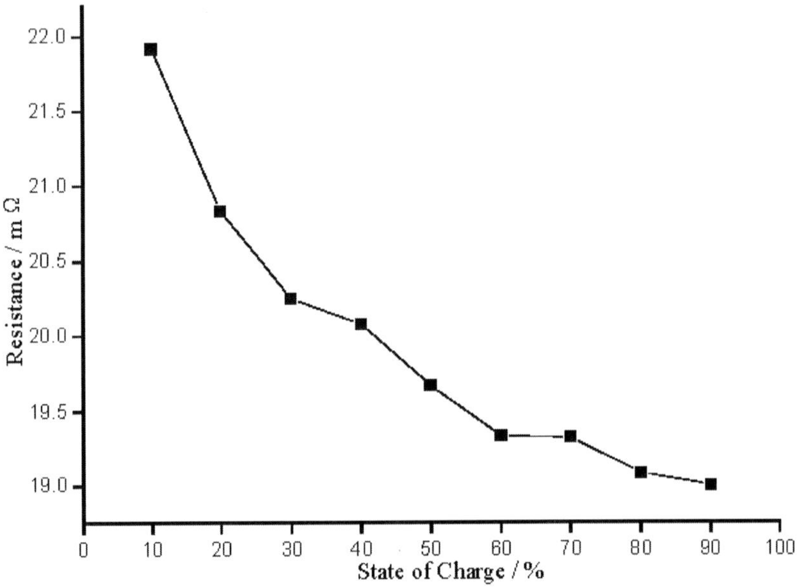

Figure 17. Resistance as a function of SOC

ii. Entropy change

Regarding entropy change, there have been many investigations and it is another factor contributing to heat generation in the energy balance. In general, the entropy change ΔS can be estimated from the temperature gradient of OCV, dV0 /dT. dV0 /dT is proportional to the entropy of reaction by the relation dV_0 /dT=ΔS /nF, where n is the number of electrons passed in the reaction and F is Faraday's constant, and reflects the influence of the incremental lithium atoms on the ordering of lithium-ion in the host lattice [18].

The entropy change of the lithium-ion battery is positive in the SOC regions of 30%-90%. This means that the thermal behavior of the battery is mainly endothermic during the discharge cycle even if there is no over-potential loss. Different entropy change under different temperature cycle is resulted from the phase change of active material, as shown in Figure 18.

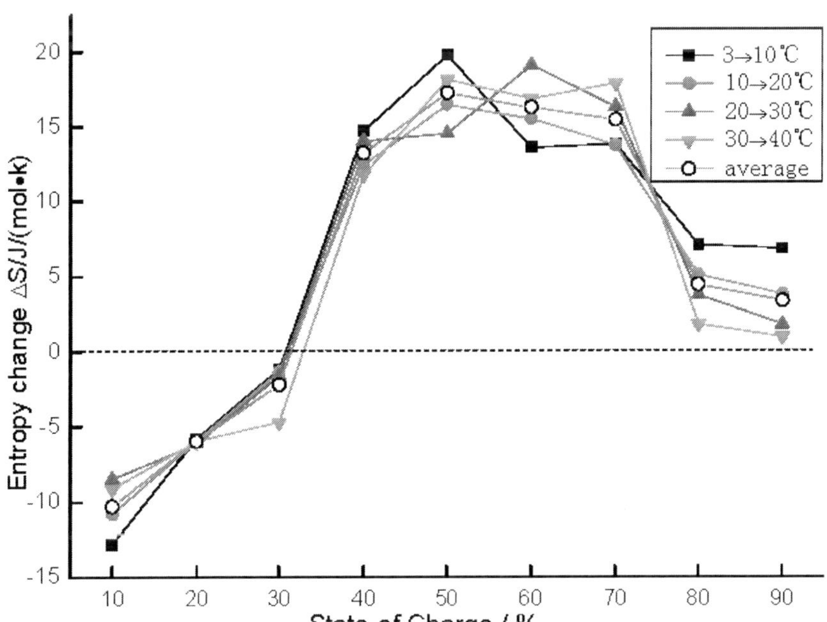

Figure 18. Entropy change for SOC and temperature

iii.Energy efficiency

The battery energy efficiency under the process of shallow charge/discharge is influenced by two factors: (1) External influence factor, such as the DOD, the ambient temperature, and the battery temperature. (2) The energy loss caused by the self-performance, for example, the increasing of resistance, the change of over-potential and the entropy. The above factors cause the energy loss and eventually reduce the battery energy efficiency.

In order to evaluate the contribution of thermal characteristics to the energy efficiency, we explore the effect under discharge and charge process by plotting the energy efficiency as a function of the DOD as shown in Figure 19. By examining the energy loss, we can determine the influence of the heat generation on the battery energy efficiency. The effects of the DOD, current rate, and temperature on energy loss are explored under different discharging interval. The results are shown in Figure 19.

The result of correlation analysis shows that there is a parabola relationship between energy efficiency and depth of discharge under the constant current rate. In addition, the lower the current rate, the higher the energy efficiency. It should be pointed out that the intercept for the minimum energy efficiency at 60% DOD reaches 85.7% at 1C. This may be related to the active material utilization at the above DOD (SOC 20%-80%).

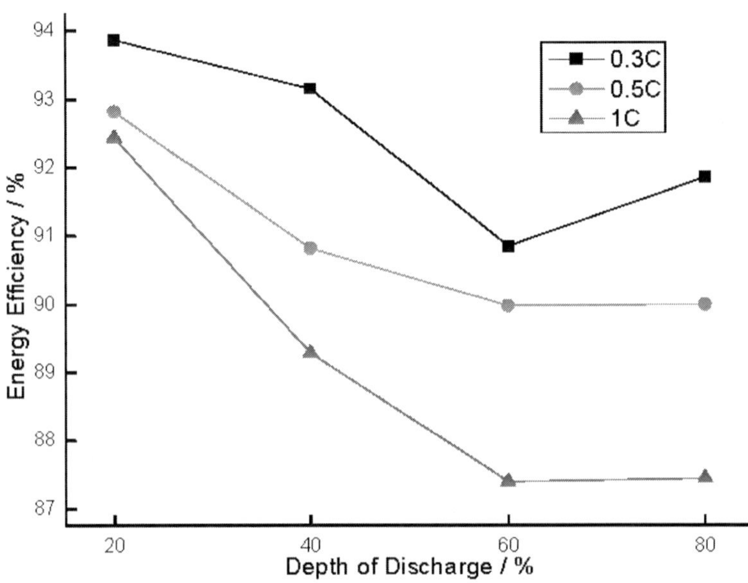

Figure 19. Energy efficiency versus DOD for lithium-ion cells

5. ADVANCED REGENERATIVE BRAKING SYSTEM (RBS) OF EVS

5.1. Introduction of RBS

RBS converts part of the mechanical energy into electric energy during braking. The drive motor functions as a generator to provide the brake torque when the vehicle brakes. The potential energy, e.g., the kinetic energy, is converted into electric energy that is then stored in an energy storage system (ESS), e.g., power battery pack, supercapacitor or their combination. When a vehicle starts or accelerates, the drive motor receives the electric energy from ESS to propel vehicle. Utilizing such energy consumption-recycling mechanism, EV recycles a significant amount of energy that is consumed in conventional vehicles [21]. The total energy efficiency and the drive range are thus improved. Recent studies indicate that the energy savings are completed from 8% to 25% of the total energy use [22-24], depending on the drive cycles and RBS performance. Additionally, the electric brake force generated from the motor is superior to the frictional brake force from the hydraulic brake actuator in terms of precision, accuracy and bandwidth for the torque control and measurement [25].

Maintaining brake safety under various braking conditions is the primary consideration when designing a vehicle brake system. The commercialized EVs/HEVs that utilize RBS still reserve the conventional hydraulic brake device in case of possible failure of the RBS. Another reason for employing the hydraulic brake is the rechargeable capacity of power battery pack. For example, when the SOC is high, the regenerative electric energy may not be fed into the battery, and thus the regenerative brake does not function. In this case, the conventional hydraulic brake is required to provide the brake force alone. The brake system with both the hydraulic brake and regenerative brake is the HBS, which has two architectures, i.e., parallel and series architectures [26]. In parallel braking architecture, both the regenerative torque and the friction torque are exerted on the same drive axle directly, while series braking allows independent modulation of the hydraulic brake torque of each axle according to the regenerative torque, and thus more kinetic energy can be recovered potentially than that with parallel braking.

Controlling the RBS involves multiple objectives, e.g., the total efficiency, brake safety, and comfort. As a RBS is an uncertainty system with parameter perturbation, strong nonlinearity and uncertain external disturbances, the classical control methods that adopt the rule based strategies [27, 28] are ineffective to guarantee the performance. Hence, researchers have studied and proposed advanced methodologies on the RBS control in the literature recently, including the fuzzy control strategies [29, 30], H∞ control [31], and neural network approaches [32]. The studies mostly focused on improving the energy recovery performance based on common factors. However, the vehicle drive safety and ride comfort, which are strongly influenced by the braking performance due to various uncontrollable phenomena by the driver's brake pedal operations [33], need sufficient considerations in the RBS control.

5.2. Regenerative Braking Control Design Based on Fuzzy Control (RBS-FC)

We propose a control methodology based on fuzzy logic for series RBS architecture to facilitate EV's energy-savings and brake safety. The RBS control aims at: 1) preventing wheel's excessive slip and lock via brake torque distribution between the front and rear axles; 2) achieving nearly the optimal regenerative efficiency under normal brake condition.

5.2.1. Braking Force Distribution

The total vehicle brake force *Fcar* that is estimated via vehicle acceleration *αcar*, consists of the front brake force *Ff* and the rear brake force *Fr*exerted on the front and rear wheels, respectively, which is shown in (17) and the first subplot of Figure 20, where *M* is the total vehicle mass. However, the load movement during braking may induce wheel lock and the consequent vehicle instability using average torque distribution between the front and rear wheels. Therefore, RBS needs appropriate brake torque distribution law to ensure the braking safety.

$$F_{car} = M \cdot \alpha_{car} = F_f + F_r \tag{17}$$

The ideal distribution curve is represented in (18) and the second subplot of Figure 20, where the symbols are indicated in the first subplot of Figure 20. When RBS control distributes the brake forces to the front and rear axles according to the ideal curve at each friction coefficient, the front and rear wheels do not lock simultaneously. The vehicle handling and stability are thus maintained.

$$F_r = \frac{1}{2}\left[\frac{G}{h_g}\sqrt{b^2 + \frac{4h_g L}{G}F_f} - \left(\frac{Gb}{h_g} + 2F_f\right)\right] \tag{18}$$

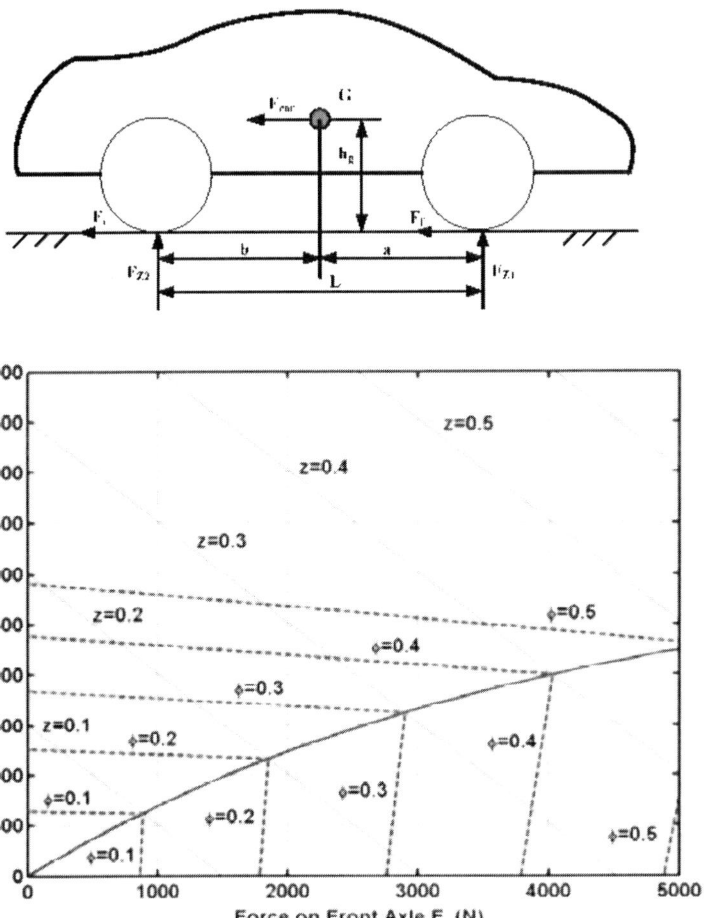

Figure 20. Vehicle braking dynamics on flat road and the ideal distribution curve

5.2.2. RBS based on Fuzzy Control (RBS-FC)

Vehicle speed and the driver's brake force command have large impacts on brake safety. The battery limitations including the battery capacity and the maximum permissible charging current protect the battery cell from damage. The capacity of the batteries is indicated with battery SOC. The maximum permissible charging current is a function of battery capacity Q, battery temperature T, battery SOC, and battery SOH, i.e., $I=f(Q,T,SOC,SOH)$. SOH is hard to determine and has little direct impact on calculating the maximum permissible charging current. Therefore, we only take SOC and battery temperature into consideration here.

The RBS-FC that uses a Takagi-Sugeno model is shown in Figure 21. There are four inputs including battery temperature, SOC, vehicle speed and driver's brake force command. The output is the regenerative brake force. The controller obtains the driver's brake force command from the pedal sensor, the vehicle speed from onboard sensor or speed estimator, and the battery SOC and temperature from the BMS. The ideal braking force distribution law gives the front brake force and the rear brake force in real time, respectively. Then, the controller determines the desired regenerative brake force according to the fuzzy rules.

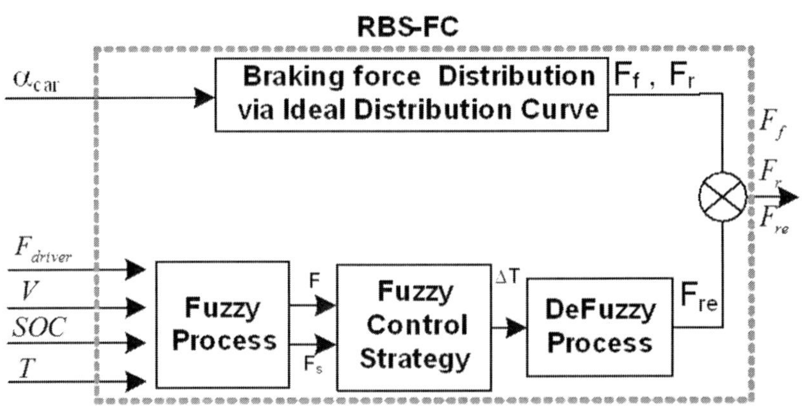

Figure 21. Structure of the control strategy system

5.2.3. Results

The proposed RBS-FC was verified in an EV prototype [22] (see Figure 22). In the test, we drove the EV on an urban road without and with RBS-FC function, respectively. The comparative results are shown in Figure 23 and Table 2, which verify that both the component efficiency and the maximum driving range are improved significantly with the proposed RBS-FC compared to the non-RBS-FC operation. The overall energy efficiency is improved from 0.341 to 0.417, i.e., an improvement of about 22%. The immediate benefit is that the maximum driving range of LF620 EV is extended from 163 kilometers to 205 kilometers for a single charge, i.e., a 26% improvement.

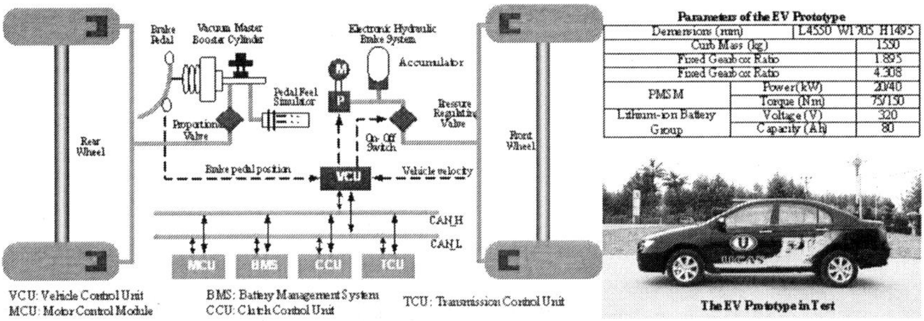

Figure 22. The test EV prototype

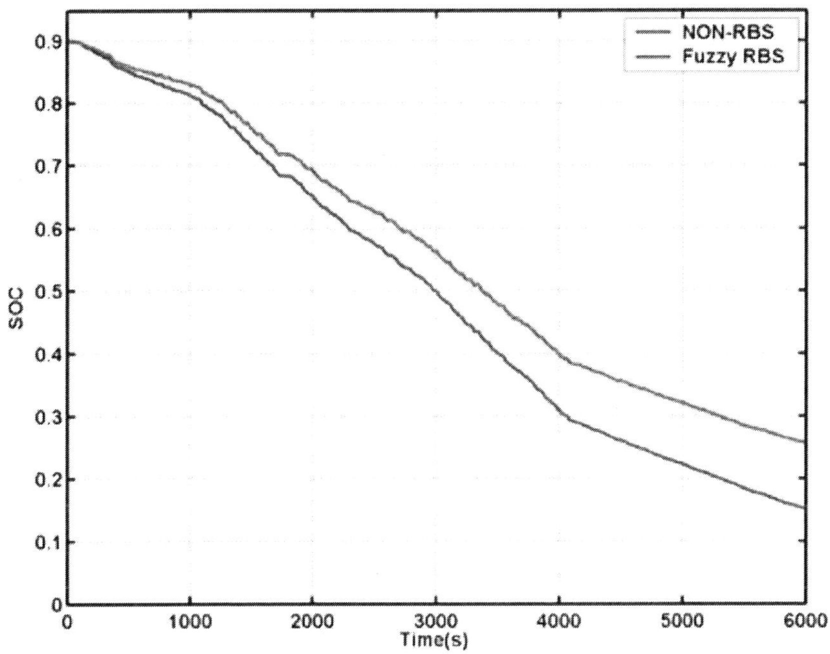

Figure 23. Comparative result of the SOC

Table 2. Comparison of the experimental results

	WITHOUT RBS-FC	WITH RBS-FC	Improvement
Motor/Controller Efficiency	0.67	0.78	16.4%
Overall Energy Efficiency	0.341	0.417	22.2%
Maximum Driving Range (km)	163	205	25.7%

6. EV ENERGY MANAGEMENT AND OPTIMIZATION

6.1. Introduction of EV energy management

The energy management here is mainly targeted at the hybrid vehicles. Hybrid vehicles use two or more than two kinds of power sources, thus the energy management strategy of hybrid vehicles is one of the most important and popular research topics, as it determines the power split between power sources and because it is related to the fuel economy of the hybrid vehicles. Several types of energy management strategies for hybrid vehicles have been developed during last few decades. These strategies can be divided into two major groups: one is on the basis of the heuristic concept; the other is based on the optimal control theory. The former mainly includes rule-based algorithms and fuzzy logic algorithms [2, 34]. Earlier in the development of hybrid vehicles, energy management strategies were dominated by these types of strategies owing to their simplicity when actually realizing them. These types of strategies, however, cannot guarantee the optimal power distribution between power sources and the optimal fuel economy as well. In addition, the rules and fuzzy logics need expert knowledge. To remedy this problem, the optimal control theory was introduced as part of the energy management strategy of hybrid vehicles, including both Dynamic Programming (DP) as developed by R. E. Bellman [35] and Pontryagin's Minimum Principle (PMP) [36-38]. The DP approach examines all admissible control inputs at every state, thus guaranteeing global optimality if the driving cycle information is given in advance [35]. However, the DP approach cannot be used directly for the real-time control of hybrid vehicles due to the backward-looking calculation process and the long calculation time. Being confronted with the drawbacks of the DP, some researchers have proposed stochastic dynamic programming (SDP) [39, 40] to overcome these problems. The PMP-based energy management strategy optimizes the power distribution between power sources and minimizes the performance measure by instantaneously providing the necessary optimality conditions. One of the major advantages of the PMP-based strategy is that there is usually one parameter to be tuned in this strategy in order to obtain optimal results over a specific driving cycle [36]. Moreover, the core of this strategy is implementable in a real-time controller, even if the driving cycle information is not known in advance [37]. Furthermore, previous research [41] proved from a mathematical point of view that the PMP-based energy management strategy can serve as a global optimal solution (DP) under the assumption that the OCV and the internal resistance of a battery are independent of the battery SOC. This assumption is reasonable for charge-sustaining types of hybrid vehicles, especially for those which use lithium-ion batteries.

6.2. Energy management based on optimal control theory

In this sub-section, two types of optimal control theory-based energy management strategies of hybrid vehicles are introduced which are the DP approach and the PMP-based strategy, respectively. The simulation results regarding the two energy management strategies are also presented.

6.2.1. The vehicle model

An FCHV is selected as an example to introduce the two energy management strategies. Figure 24 [42] shows the block diagram of an FCHV in which an FCS and a battery are the two power sources. The electrical energy provided by the FCS and the battery is converted to the mechanical energy through the electric motor to propel the vehicle. Meanwhile, the electric motor can be controlled to work as a generator to convert a part of braking energy of the vehicle to electrical energy and store it into the battery. A group of vehicle parameters are selected by consulting the available literature [43] for the simulation study as listed in Table 3. The powertrain information is also summarized in Table 3. Details on this vehicle model can be found in the literature [42, 44, 45].

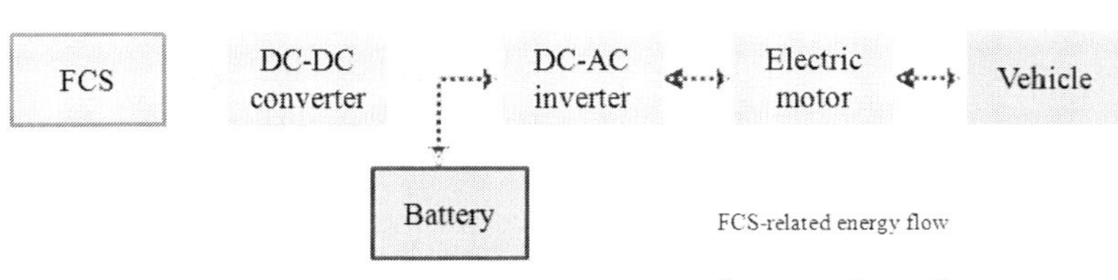

Figure 24. Configuration and energy flow of an FCHV

Table 3. Vehicle parameters and powertrain information of the FCHV

Parameter	Value
Vehicle total mass (kg)	1500
Final drive gear efficiency (%)	95
Tire radius (m)	0.29
Aerodynamic drag coefficient	0.37
Vehicle frontal area (m^2)	2.59
Air density (kg/m^3)	1.21
Rolling resistance coefficient	0.014
Electric motor (kW)	75
FCS (kW)	45
Battery (kWh)	1.9
Efficiency of converter and inverter (%)	95

6.2.2. DP approach

The DP examines all possible control inputs at every state [46]. Thus, the optimal trajectory to the final state can be obtained for every state in the DP. The DP needs much more calculation time compared to the PMP due to above characteristics.

The objective of the optimal control problem here is to find an optimal power split ratio trajectory during driving which brings the minimum fuel consumption. Here, we define the battery power P_{bat} as the control variable and the battery SOC as the state variable. The state equation of the optimal control problem expresses the dynamic behavior of the battery as shown in (19), in which Q_{bat} is the capacity and V is the open circuit voltage and R is the internal resistance of the battery.

$$\dot{SOC} = -\frac{1}{Q_{bat}} \frac{V(SOC) - \sqrt{V(SOC)^2 - 4R(SOC) \cdot P_{bat}}}{2R(SOC)}$$

$$\dot{SOC} = f(P_{bat}, SOC) \qquad (19)$$

Now, we rewrite (19) in a discrete form as (20):

$$x(k+1) = x(k) + \Delta t \cdot f(u, x) \qquad k = 0, 1, \cdots, N-1 \qquad (20)$$

Here, x is the state variable and u is the control variable. N is the number of time steps and f is the same function as in (19).

The performance measure J to be minimized here is the total fuel consumption. Optimal performance measure from every possible state at every time step to the designated final state can be expressed as follows:

$$J^*_{N-k,N}(x(N-k)) = \min_{u(N-k)} (\Delta t \cdot \dot{m}_{h_2}(x(N-k), u(N-k)) + J^*_{N-(k-1),N}(x(N-(k-1))))$$
$$x(N-(k-1)) = x(N-k) + \Delta t \cdot f(u(N-k), x(N-k))$$
$$k = 0, 1, \cdots, N-1 \quad (21)$$

Here, $mh2$·is the fuel consumption rate of the FCS. $J*N-k,N$ represents the optimal performance measure from $N-k$ time step to the final time step N for every possible state at $N-k$ time step. The DP examines all possible control variables at every state and selects the optimal one to every state. $J*N-(k-1),N$ represents the optimal performance measure from $N-(k-1)$ time step to the final time step N for every possible state at $N-(k-1)$ time step.

Due to the calculation characteristics of the DP, the optimal results can be acquired only when the DP calculation is finished. The results of the DP form a field which is defined as the optimal field here. The optimal field includes not only the optimal state trajectory information but also the optimal performance measure information. Figure 25 illustrates an example of the optimal field of the FTP72 urban driving cycle. Here, the initial battery SOC and final SOC are all set to 0.6, and the line on the optimal field indicates the optimal trajectory of the battery SOC.

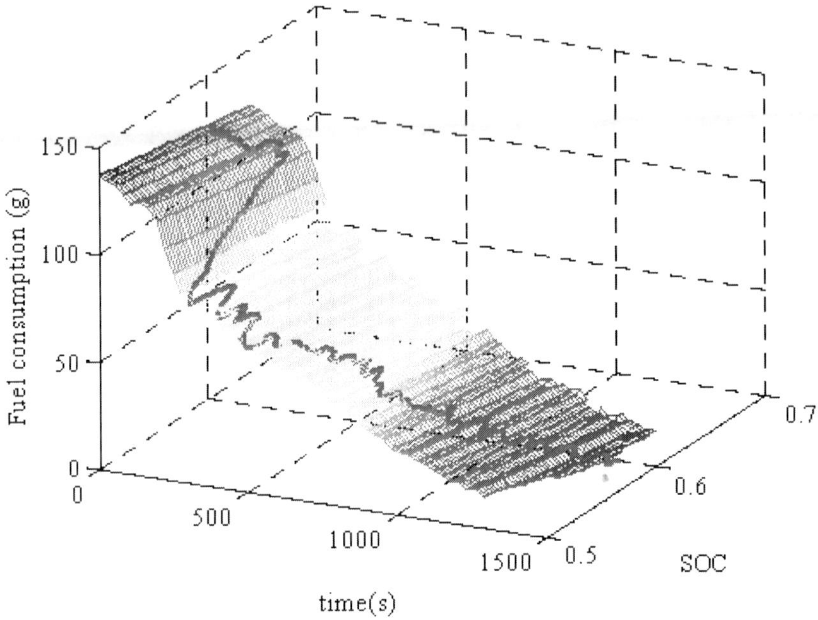

Figure 25.The optimal field of the FTP72 urban driving cycle

6.2.3. PMP-based strategy

The PMP-based energy management strategy optimizes the power distribution between the FCS and the battery instantaneously by providing the necessary optimality conditions. Here, we define the FCS net power P_{fcs} as the control variable, whereas the control variable in 5.2.2 is the battery power P_{bat}. The state equation of the control system is given in equation (19). Here, the time derivative of the battery SOC is expressed by P_{fcs} and the battery SOC instead of P_{bat} and SOC, as the total power requirement of the vehicle can be obtained when a driving cycle is given in advance. Therefore, the state equation of the optimal control problem is transformed using another function F, as follows:

$$\dot{SOC} = F(P_{fcs}, SOC) \tag{22}$$

Considering the state equation (22), which is also a constraint on the optimal control problem, and the relationship between P_{fcs} and m_{h2}·, the performance measure is as follows:

$$J = \int_{t_0}^{t_f} \left\{ \dot{m}_{h_2}(P_{fcs}) + p \cdot \left(F - \dot{SOC} \right) \right\} dt \tag{23}$$

Here, p is the Lagrange multiplier, which is also defined as the co-state in the PMP and is an equivalent parameter between the fuel usage and the electric usage.

According to the optimal control theory based on the Calculus of Variation, the necessary conditions of the optimal control problem can be obtained when the variation of the performance measure δJ is zero [46]. If we introduce a Hamiltonian H, which is defined as

$$H = \dot{m}_{h_2}(P_{fcs}) + p \cdot F(P_{fcs}, SOC), \tag{24}$$

then the necessary conditions that minimize the performance measure (23) from time $t0$ to time tf are as follows:

$$\frac{\partial H}{\partial p} = \dot{SOC}$$
$$\frac{\partial H}{\partial SOC} = -\dot{p}$$
$$\frac{\partial H}{\partial P_{fcs}} = 0 \tag{25}$$

The first necessary condition in (25) is the state equation. The second necessary condition is the co-state equation, which determines the optimal trajectory of the co-state p. The third necessary condition determines the optimal trajectory of the control variable P_{fcs} by minimizing the Hamiltonian H.

The PMP is the general case of the Calculus of Variation, in which the third necessary condition in (25) is expressed as follows:

$$H(P_{fcs}^{*}, SOC^{*}, p^{*}) \leq H(P_{fcs}, SOC^{*}, p^{*}) \tag{26}$$

The advantage of this form is that it can be applied to non-linear or non-differentiable or non-convex function [41].

Figure 26 [47] illustrates an example of solving the optimal solution according to the PMP. From the shape of the third subplot, it can be observed that the optimal FCS net power can be obtained at this moment, which minimizes the Hamiltonian. The similar process is repeated at every calculation time step, and the optimal solution trajectory can be obtained finally.

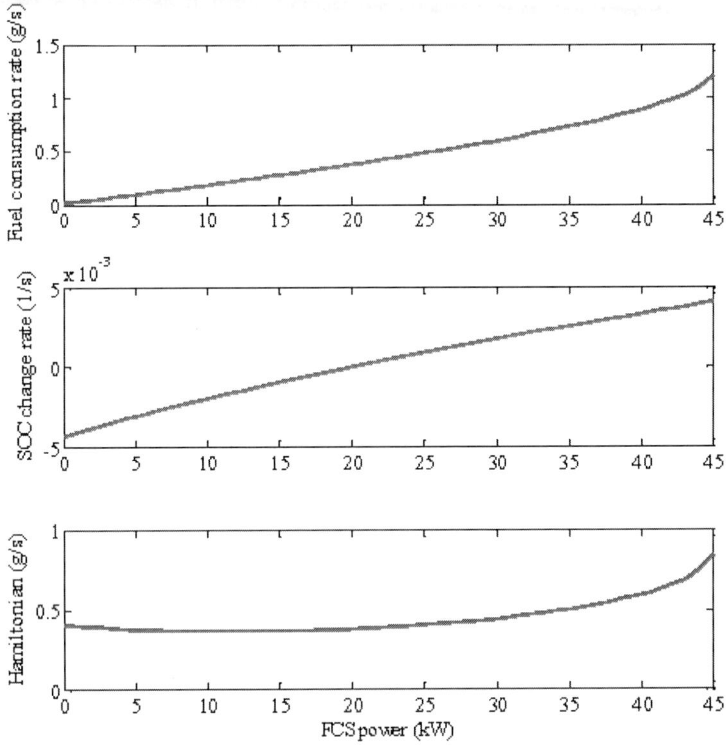

Figure 26. An example of solving the optimal FCS net power according to the PMP

Table 4. Fuel economy comparison between DP and PMP

	FTP72 urban	NEDC 2000	Japan 1015
DP (kg/100 km)	1.063	1.195	1.012
PMP (kg/100 km)	1.064	1.195	1.017
Discrepancy (%)	0.094	0	0.494

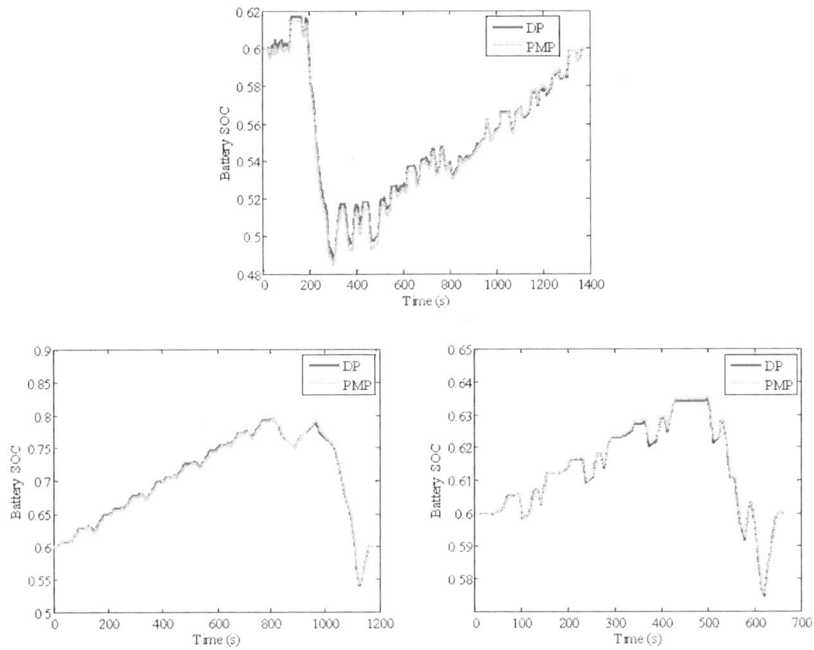

Figure 27. Comparison of DP and PMP on the optimal battery SOC trajectories

6.2.4. Comparison of PMP-based strategy and DP approach

Simulation results of the above two energy management strategies are compared on three typical driving cycles here. The first subplot in Figure 27 [48] illustrates the optimal battery SOC trajectories on the FTP72 urban driving cycle solved by the PMP and the DP. As we can see, the two trajectories overlap each other most of the time. However, the two trajectories are not perfectly the same. Different calculation environment for each energy management strategy may have influence on this fact. The trajectories will be closer to each other if the same calculation environment is used. The second and the third subplots in Figure 27 are for the cases of the NEDC 2000 and the Japan 1015 driving cycle, respectively. Table 4 [48] shows the fuel economy comparison of the two energy management strategies. As we can see, the discrepancy between the two strategies is within 0.5 % for the three driving cycles.

6.3. Energy management using prediction of future driving condition

In 5.2, the energy management strategies are evaluated on some specific driving cycles. However, the driving cycle is not fixed in the real-world driving. Currently, the vehicle telemetry, such as the Global Positioning System (GPS) and the Intelligent Transportation System (ITS), is actively used to vehicles in order to provide traffic preview information to the drivers. This information can be used when constructing energy management strategies of hybrid vehicles. In this section, the PMP-based energy management strategy introduced in 5.2.3 is extended in order to optimize the power distribution and as well as the vehicle driving route based on the traffic preview information.

The control objective here is to optimize the vehicle driving route and the power split ratio at the same time using the traffic preview information, so that the fuel consumption is further minimized. There are two state variables in this control problem, which are the battery SOC and the vehicle velocity v_{veh}. The two state equations are as follows:

$$\dot{SOC} = f\left(SOC, P_{fcs}, v_{veh}, a_{veh}\right)$$
$$\dot{v}_{veh} = a_{veh} \tag{27}$$

The first equation in (27) is the dynamic behavior of the battery, and the second equation represents the vehicle dynamic behavior. Here, the FCS net power P_{fcs} and the vehicle acceleration a_{veh} are two control variables of the control problem.

Constraints on the state variables and on the control variables are as follows:

$$SOC_{\min} \leq SOC(t) \leq SOC_{\max}$$
$$v_{\min}(t) \leq v_{veh}(t) \leq v_{\max}(t)$$
$$P_{fcs,\min}(t) \leq P_{fcs}(t) \leq P_{fcs,\max}(t)$$
$$a_{\min}(t) \leq a_{veh}(t) \leq a_{\max}(t)$$
$$\int_{t_0}^{t_f} v_{veh}(t)dt = L \tag{28}$$

Here, $t0$ and tf are the initial time and final time of the preview length, and L is the preview length. L, v_{min}, and v_{max} can be obtained from the GPS and ITS devices. The battery SOC constraint is time-invariant, and the maximum and minimum values of other cases are time-variant.

The application process of the PMP for the two-state variable system is similar to that for a single state variable system. According to the PMP, when a Hamiltonian is defined as follows

$$H\left(SOC, P_{fcs}, v_{veh}, a_{veh}\right) = \dot{m}_{h_2}\left(P_{fcs}\right) + p_1 \cdot f\left(SOC, P_{fcs}, v_{veh}, a_{veh}\right) + p_2 \cdot a_{veh}, \quad (29)$$

the necessary conditions that can achieve the goal of this control problem are as follows:

$$\begin{aligned}
\dot{SOC}^* &= \frac{\partial H}{\partial p_1}\left(SOC^*, P_{fcs}^*, v_{veh}^*, a_{veh}^*\right) \\
\dot{v}_{veh}^* &= \frac{\partial H}{\partial p_2}\left(SOC^*, P_{fcs}^*, v_{veh}^*, a_{veh}^*\right) \\
\dot{p}_1^* &= -\frac{\partial H}{\partial SOC}\left(SOC^*, P_{fcs}^*, v_{veh}^*, a_{veh}^*\right) \\
\dot{p}_2^* &= -\frac{\partial H}{\partial v_{veh}}\left(SOC^*, P_{fcs}^*, v_{veh}^*, a_{veh}^*\right) \\
H\left(SOC^*, P_{fcs}^*, v_{veh}^*, a_{veh}^*, p_1^*, p_2^*\right) &\le H\left(SOC^*, P_{fcs}, v_{veh}^*, a_{veh}, p_1^*, p_2^*\right)
\end{aligned} \quad (30)$$

Here, the first two necessary conditions are the two state equations, which are also constraints on the optimal control problem. $p1$ and $p2$ are two co-states which can be determined based on the third and the fourth necessary conditions. At every calculation time step, the optimal FCS net power and the optimal vehicle acceleration can be found by the fifth necessary condition.

Figure 28 [49] illustrates an example of simulation results. A FCHV with the total mass of 1700kg is used in this simulation. This figure shows the maximum and the minimum vehicle velocities along the preview driving which are assumed to be provided by the GPS and ITS devices. This figure also shows a benchmark vehicle velocity profile, which is within the above two velocity profiles, and the optimal vehicle velocity profile derived from the proposed energy management strategy. The length discrepancy between the optimal and the benchmark vehicle velocity profiles is around three meters here. The fuel economy is improved through the proposed energy management strategy compared to the benchmark case. In the benchmark case, the vehicle velocity profile is fixed and only the power split ratio is optimized. Table 5 [49] shows the fuel consumption information of the two vehicle velocity profiles. This table reflects that the fuel consumption is improved around 10.4 % by the proposed energy management strategy.

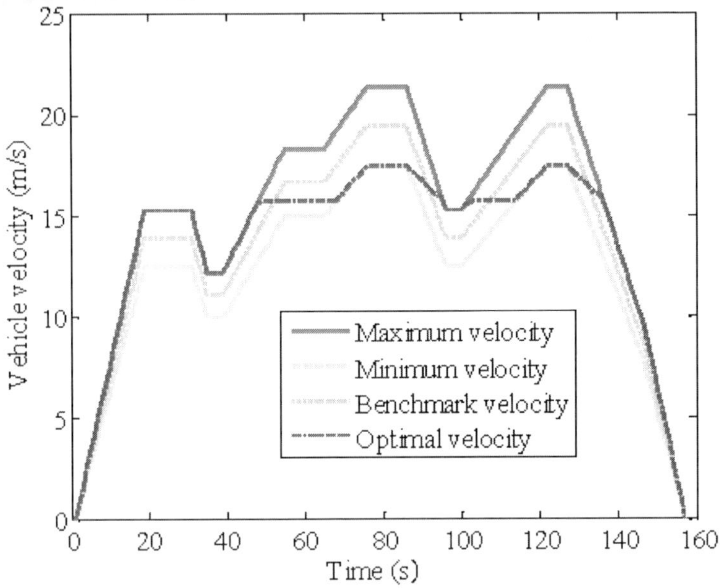

Figure 28. The optimal vehicle velocity profile and the benchmark velocity profile

Table 5. Fuel consumption of the optimal vehicle velocity profile and the benchmark velocity profile

	Fuel consumption (kg/100km)	Driving distance (m)
Benchmark velocity profile	1.030	2173
Optimal velocity profile	0.923	2170

7. CONCLUSIONS

This chapter deals with the energy processing and control of EVs. The latest energy saving and optimal control technologies are introduced and the key results contributed by the authors on the high-efficiency traction motor, intelligent battery management system, regenerative braking with high energy recovery, and the vehicle energy management are presented. The following points are drawn from this chapter:

1. For the integrated motor drive study, the integrated SRM drive obtains the battery charger and grid-connected inverter functions, whereas it saves the cost, space, and weight. The TDM method is used to improve the coefficient of utilization of the components. The suitable parking position is discussed for the stable working in the bi-directional mode. The power factor correction is considered and well worked in the integrated SRM drive.
2. For the BMS study, the proposed quantitative relationship exhibits the functions of the terminal voltage, resting time, and temperature to the initial battery SOC. This achievement raises the accuracy of the battery SOC estimation system and improves match flexibility between the battery SOC and the power output. The correlation analysis result of the energy efficiency shows that there is a parabola relationship between the energy efficiency and the DOD under the constant current rate. The battery energy conversion efficiency is less influenced by the thermal effect under the smaller current rate.
3. For the RBS study, the comparative results verify that both the component efficiency and the maximum driving range are improved significantly with the proposed RBS-FC compared to the non-RBS-FC operation. The overall energy efficiency is improved around 22%. The immediate benefit is that the maximum driving range of LF620 EV is extended from 163 kilometers to 205 kilometers for a single charge, i.e., a 26% improvement.
4. For the energy management and optimization study, the PMP-based strategy obtains almost the same optimal trajectories with the DP approach, while it saves much more computation time compared to the DP approach. The extended PMP-based strategy, which optimizes the power distribution and as well as the vehicle driving route based on the traffic preview information, improves the fuel economy around 10.4 % compared to a benchmark case studied in this chapter.

8. ACKNOWLEDGEMENTS

This work was supported by National Natural Science Foundation of China (Grant No. 51305437 and No. 51107142), Shenzhen Basic Research Fund (JCYJ20130401170306801, JCYJ20120617121836364, JCYJ201304011 70306854), and Guangdong Innovative Research Team Program (No. 201001D0104648280).

REFERENCES

1. Chan CC, Bouscayrol A, Chen K. Electric, Hybrid, and Fuel-cell Vehicles: Architectures and Modeling. IEEE Transactions on Vehicular technology 2010; 59(2) 589-598.
2. Ehsani M, Gao Y, Emadi A. Modern Electric, Hybrid Electric, and Fuel Cell Vehicles. Boca Raton: CRC Press; 2010.
3. Chan CC. The State of the Art of Electric, Hybrid, and Fuel Cell Vehicles. Proceeding of the IEEE 2007; 95(4) 704-718.
4. Dyke KJ, Schofield N, Barnes M. The impact of transport electrification on electrical networks. IEEE Trans. Ind. Electron. 2010; 57(12) 3917-3926.
5. Mapelli FL, Tarsitano D, Mauri M. Plug-in hybrid electric vehicle: modeling, prototype, realization, and inverter losses reduction analysis. IEEE Trans. Ind. Electron. 2010; 57(2) 598-607.
6. Saber AY, Venayagamoorthy GK. Plug-in vehicles and renewable energy sources for cost and emission reductions. IEEE Trans. Ind. Electron. 2011; 58(4) 1229-1238.
7. Wieringen V, Pop-Iliev R. Development of a dual-fuel power generation system for an extended range plug-in hybrid electric vehicle. IEEE Trans. Ind. Electron. 2010; 57(2) 641-648.
8. Sul SK, Lee SJ. An integral battery charger for four wheel drive electric vehicle. IEEE Trans. Ind. Appl. 1995; 31(5) 1096-1099.
9. Pellegrino G, Armando E, Guglielmi P. An integral battery charger with power factor correction for electric scooter. IEEE Trans. Pow. Electron. 2010; 25(3) 751-759.
10. Liang J, Lee DH, Xu G, Ahn JW. Analysis of passive boost power converter for three-phase SR drive. IEEE Transactions on Industrial Electronics 2010; 57(9) 2961-2971.
11. Chang HC, Liaw CM. Development of a compact switched-reluctance motor drive for EV propulsion with voltage-boosting and PFC charging capabilities. IEEE Transactions on Vehicular Technology 2009; 58(7) 3198-3215.
12. Liang J, Xu G, Wang B, Wang H. A Novel Integrated Switched Reluctance Motor Drive with Bi-directional Inverter: proceedings of IEEE International Conference on Industrial Technology, 26 Feb.-1 Mar. 2014, Busan, Korea.

13. Liang J, Xu G, Jian L, Chang M. Analysis Parking Position of Integrated Switched Reluctance Motor Drive System with On-board Charger for EV: proceedings of the 8th IEEE Vehicle Power and Propulsion Conference, 9-12, Oct. 2012, Seoul, Korea.

14. Li JH, Barillas JK, Guenther C, Danzer MA. A Comparative Study of State of Charge Estimation Algorithms for LiFePO4 Batteries Used in Electric Vehicles. Journal of Power Sources 2013; 230 244-250.

15. Zhang YH, Song WJ, Lin SL, Lv J, Feng ZP. A Critical Review on State of Charge of Batteries. Journal of Renewable and Sustainable Energy 2013; 5(2) 021403-1-021403-10.

16. He HW, Zhang XW, Xiong R, Xu Y, Guo Hq. Online Model-Based Estimation of State of Charge and Open Circuit Voltage of Lithium Ion Batteries in Electric Vehicles. Energy 2012; 39(1) 310-318.

17. He HW, Xiong R, Guo HQ. Online Estimation of Model Parameters and State-of-Charge of LiFePO4 Batteries in Electric Vehicles. Applied Energy 2012; 89(1) 413-420.

18. Thomas KE, Bogatu C, Newman J. Measurement of the Entropy of Reaction as a Function of State of Charge in Doped and Undoped Lithium Manganese Oxide. Journal of the Electrochemical Society 2001; 148(6) A570-A575.

19. Onda K, Kameyama H, Hanamoto T, Ito K. Experimental Study on Heat Generation Behavior of Small Lithium-Ion Secondary Batteries. Journal of the Electrochemical Society 2003; 150(3) A285-A291.

20. Bandhauer TM, Garimella S, Fuller TF. A Critical Review of Thermal Issues in Lithium-Ion Batteries. Journal of the Electrochemical Society 2011; 158(3) R1-R25.

21. Gao YM, Chen LP, Ehsani M. Investigation of the effectiveness of regenerative braking for EV and HEV. SAE Trans 1999; 108 3184-3190.

22. Xu GQ, Li WM, et al. An Intelligent Regenerative Braking Strategy for Electric Vehicles. Energies 2011; 4(9) 1461-1477.

23. Yao J, Zhong ZM, Sun ZC. A fuzzy logic based regenerative braking regulation for a fuel cell bus: proceedings of the IEEE International Conference on Vehicular Electronics and Safety, 13-15 December 2006, Beijing, China.

24. Zolot M, Markel T, Pesaran A. Analysis of fuel cell vehicle hybridization and implications for energy storage devices: proceedings of the 4th

Advanced Automotive Battery Conference, 2-4 June 2004, San Francisco, CA, USA.

25. Yin GD, Jin XJ. Cooperative Control of Regenerative Braking and Antilock Braking for a Hybrid Electric Vehicle. Mathematical Problems in Engineering 2013; 1-9.

26. Guo HQ, He HW, et al. A Combined Cooperative Braking Model with a Predictive Control Strategy in an Electric Vehicle. Energies 2013; 6(12) 6455-6475.

27. Zhang CW, Bai ZF, Cao BG, Li JC. Study on regenerative braking of electric vehicle: proceedings of the 4th International Power Electronics and Motion Control Conference, 14-16 August 2004, Xi'an, China.

28. Guo JG, Wang JP, Cao BG. Regenerative braking strategy for electric vehicles: proceeding of the 2009 IEEE Intelligent Vehicles Symposium, 3-5 June 2005, Xi'an, China.

29. Zhang JM, Song BY, Cui SM. Fuzzy logic approach to regenerative braking system: proceedings of the International Conference on Intelligent Human-Machine Systems and Cybernetics, 26-27 August 2009, Hangzhou, China.

30. Paterson J, Ramsay M. Electric vehicle braking by fuzzy logic control: proceedings of the IEEE Industry Applications Society Annual Meeting, 2-8 October 1993, Toronto, Canada.

31. Ye M, Bai Z, Cao B. Robust Control for Regenerative Braking of Battery Electric Vehicle. IET Control Theory and Applications 2008; 2(12) 1105-1114.

32. Gao HW, Gao YM, Ehsani M. A neural network based SRM drive control strategy for regenerative braking in EV and HEV: proceedings of the Electric Machines and Drives Conference, 17-20 June 2001, Cambridge, MA, USA.

33. Mutoh, N, Hayano Y, Yahagi H, Takita K. Electric braking control methods for electric vehicles with independently driven front and rear wheels. IEEE Transactions on Industrial Electronics. 2007; 54(2) 1168-1176.

34. Gao DW, Jin ZH, Lu QC. Energy Management Strategy Based on Fuzzy Logic for a Fuel Cell Hybrid Bus. Journal of Power Sources 2008; 185(1) 311-317.

35. Lin CC, Peng H, Grizzle JW, Kang J. Power Management Strategy for a Parallel Hybrid Electric Truck. IEEE Transactions on Control Systems Technology 2003; 11(6) 839-849.

36. Serrao L, Rizzoni G. Optimal control of power split for a hybrid electric refuse vehicle: proceedings of American Control Conference, 11-13 June 2008, Washington, USA.

37. Chasse A, Sciarretta A. Supervisory Control of Hybrid Powertrains: an Experimental Benchmark of Offline Optimization and Online Energy Management. Control Engineering Practice 2011; 19(11) 1253-1265.

38. Kim NW, Cha SW, Peng H. Optimal Equivalent Fuel Consumption for Hybrid Electric Vehicles. IEEE Transactions on Control Systems Technology 2012; 20(3) 817-825.

39. Lin CC, Peng H, Grizzle JW. A stochastic control strategy for hybrid electric vehicles: proceedings of American Control Conference, 30 June-2 July 2004, Boston, USA.

40. Kim MJ, Peng H. Power Management and Design Optimization of Fuel Cell/Battery Hybrid Vehicles. Journal of Power Sources 2007; 165(2) 819-832.

41. Kim NW, Cha SW, Peng H. Optimal Control of Hybrid Electric Vehicles Based on Pontryagin's Minimum Principle. IEEE Transactions on Control Systems Technology 2011; 19(5) 1279-1287.

42. Zheng CH, Xu GQ, Cha SW, Park YI, Lim WS. PMP-Based Power Management Strategy for Two-State Variable FCHV Systems and Its Optimality. International Journal of Precision Engineering and Manufacturing 2014; 15(4) 769-776.

43. Bernard J, Delprat S, Buechi F, Guerra TM. Fuel-Cell Hybrid Powertrain: Toward Minimization of Hydrogen Consumption. IEEE Transactions on Vehicular Technology 2009; 58(7) 3168-3176.

44. Zheng CH, Kim NW, Cha SW. Optimal Control in the Power Management of Fuel Cell Hybrid Vehicles. International Journal of Hydrogen Energy 2012; 37(1) 655-663.

45. Zheng CH, Xu GQ, Park YI, Lim WS, Cha SW. Prolonging Fuel Cell Stack Lifetime Based on Pontryagin's Minimum Principle in Fuel Cell Hybrid Vehicles and Its Economic Influence Evaluation. Journal of Power Sources 2014; 15 533-544.

46. Kirk DE. Optimal Control Theory. New York: Prentice-Hall; 1970.

47. Zheng CH, Oh CE, Park YI, Cha SW. Fuel Economy Evaluation of Fuel Cell Hybrid Vehicles Based on Equivalent Fuel Consumption. International Journal of Hydrogen Energy 2012: 37(2) 1790-1796.

48. Zheng CH, Park YI, Lim WS, Cha SW, Xu GQ. Comparison of PMP and DP in Fuel Cell Hybrid Vehicles. International Journal of Automotive Technology 2014: 15(1) 117-123.

49. Zheng CH, Xu GQ, Cha SW. A power management strategy of hybrid vehicles using traffic preview information: proceedings of Vehicular Technology Conference 2014-Spring, 18-21 May 2014, Seoul, Korea.

3
CHAPTER

INCREASING THE EFFICIENCY OF GRID TIED MICRO WIND TURBINES IN LOW WIND SPEED REGIMES

Kshitij Tiwari[1,2], Tirumalachetty Harinarayana[2]

[1]Department of Electrical Engineering, Indian Institute of Technology, Kanpur, India

[2]Gujarat Energy Research and Management Institute, Gandhinagar, India

ABSTRACT

Major problem with grid tied micro wind turbine is synchronization and wind variability. Due to this problem the stability of available grid gets reduced. The stability can be achieved by output power control of the turbine. Major part of many countries like India, the annual mean wind speed is not high. The rated wind speed of turbine remain around 11 m/s and cut in is around 3.5 m/s. Due to this problem we aimed to develop a sustainable wind energy system that can provide stable power supply even at the locations of low wind speed of 2 - 4 m/s. To address this issue, a momentary impulse or external torque to the rotor by external motor is one of the good options to maintain the momentum of blades and thus provide stability for sufficient time. Various theoretical calculations and experiments are conducted on the above method. This would increase the output power and also the efficiency of wind turbine. We show that Return-On-Investment will be high as compared with other grid connected turbines. Our proposed concept in the present study, if implemented properly, can help the installation of number of wind turbines even at domestic level. It also makes the consumers energy independent and promotes the use of wind as a source of energy and may enter as a rooftop energy supply system similar to solar.

Keywords: Grid Tied Rooftop Micro Wind Turbine, Small External DC Motor, Wind Variability, Tip-Speed Ratio (TSR), Stabilized Grid Supply

1. INTRODUCTION

Wind energy development is picked up as a part of electricity generation in different parts of India. Although it was initiated a decade back, there is still enough scope for its improvement in this technology. The reality is that our life-style in the early twenty-first century requires huge energy and wind turbines are one of the least environmentally harmful ways of generating electricity. Of the installed capacity, wind turbines have a large share among the other grid connected renewable sources, close to 66.66%[1]. Obviously wind turbines need to be built where the wind blows reliably and strongly. The society is certainly not willing to give up private cars, air conditioning, computers, television, refrigeration, and all the other energy consuming equipment of the modern life-style. Accordingly, we must build sustainable energy supply systems. Earlier works promoted wind energy in many ways [1] [2] . In our study we have worked on grid tied rooftop micro wind turbine. The major problem with such a system is synchronization largely due to wind variability. Due to this problem, the stability of available grid gets reduced. This can somehow be achieved by the output power control of the turbine. India is the fifth largest wind power producer in the world. As of March 2014, the installed capacity of wind power in India is 21136.3 MW [3] [4] , mainly spread across Tamil Nadu (7154 MW), Gujarat (3093 MW), Maharashtra (2976 MW), Karnataka (2113 MW) and Rajasthan (2355 MW). Wind power accounts for 8.5% of India's total installed renewable power capacity, and it generates 1.6% of the country's power. Still it is not very much popular at domestic level due to the high initial investment and other problems as discussed above. Our goal is to achieve stable power supply to the grid from wind source. To this issue, we suggest a new method of small external motor that will provide momentary impulse or external torque to the rotor through relay switch or voltage controller that helps to maintain the momentum of blades for sufficient time and provide stable power supply. In this method, the light weight external motor has been attached at the hub of the wind turbine and it works on the method of spring balance system. The motor aligns itself with the axis of rotation of blades when a mass of around 10 pounds is put to the groove attached to motor. This is done so as to not disturb the aerodynamics of the blades and the yaw system. The motor provides external torque to the rotor only during moderately low wind speeds, i.e., around 2.5 - 4 m/s. At other times, when the wind speed is self sufficient to produce stable power supply, the motor remains off-axis from the rotor and doesn't interfere in its motion. This mechanism is currently mechanical but could be automated for better performance. The concept presented in this paper is implemented and tested.

2. MATERIALS AND METHODS

In our method, we attached an external motor to the turbine rotor that provides sufficient external torque to the rotor to rotate at its rated rpm value when the wind speed is less than the rated speed so that the wind generator can generate a stabilized power supply.

2.1. Theory

It is well known that the amount of power extracted from the wind is given by the equation:

$$P_{\text{Wind}} = \frac{1}{2}\rho A V^3 C_p$$

[5] -[7] (1)

$$P_{\text{Generated}} = \frac{1}{2}\rho \pi R^2 V^3 C_p C_t C_g \left(\frac{p}{2}\right)$$

[5] -[7] (2)

$$\text{Tip Speed Ratio}(\lambda) = \frac{\omega R}{V}$$

[5] -[7] (3)

where, P_{Wind} = Power Delivered by Wind to the rotor below the rated value, $P_{\text{Generated}}$ = Output Power of the generator, C_p = Power Coefficient, C_t = Transmission Coefficient, C_g = Generator Efficiency, p = Number of poles in Generator.

Motor supply sufficient torque to the rotor so that generator operates at rated power. The following is the mathematical condition:

$$\tau_{\text{Wind}} + \tau_{\text{Motor}} = \tau_{\text{Rated}} \tag{4}$$

$$\frac{P_{\text{Wind}}}{\omega_{\text{Rotor}}} + \frac{P_{\text{Motor}}}{\omega_{\text{Motor}}} = \frac{P_{\text{Rated}}}{\omega_{\text{Rated}}}$$

$$\frac{\frac{1}{2}\rho\pi R^2 V_{\text{Wind}}^3 C_P}{\omega_{\text{Rotor}}} + \frac{P_{\text{Motor}}}{\omega_{\text{Motor}}} = \frac{\frac{1}{2}\rho\pi R^2 V_{\text{Rated}}^3 C_P}{\omega_{\text{Rated}}}$$

Using Equation (3)

$$P_{\text{Motor}} = \frac{1}{2}\frac{\rho\pi R^3 C_P \left(V_{\text{Rated}}^2 - V_{\text{Wind}}^2\right)\omega_{\text{Motor}}}{\lambda}$$

For the above suggested mechanism to be useful, the generated power should be greater than the power supplied by the motor:

$$P_{\text{Generated}} \geq P_{\text{Motor}}$$

$$V_{\text{Rated}}^3 C_t C_g \left(\frac{p}{2}\right) \geq \frac{R\omega_{\text{Motor}}}{\lambda}\left(V_{\text{Rated}}^2 - V_{\text{Wind}}^2\right) \tag{5}$$

In our study, SNT-1[2] 700 W micro grid tied turbine is used. Its parameters are, V_{Rated} = 11 m/s (according to industry standards), C_t = 1 (since no gearbox system is used), C_g = 0.25 (assumed to be constant for easier calculations), C_P = 0.33 (assumed to be constant), p = 8, R = 920 mm, $V_{Cut\text{-}in}$ = 2m/s, $(\omega_{Motor})_{Max}$ = 70 rpm, $(P_{Motor})_{Max}$ = 18 Watts.

SNT-1[2] turbine is although designed for maximum rotational speed of 300 rpm at a wind speed of 6 m/s. So, TSR = 4.81 (assumed as constant for easier calculations) [8] -[10] . Above the wind speed of 6 m/s, it dampens extra energy to the dummy load.

Accordingly, the calculations have been carried out considering the wind speed of 6 m/s as rated. Another reason for this is that the generator is not designed for higher wind speeds.

2.2. Theoretical Results

Following are the results observed by calculating using theoretical considerations:

1) Following are the results to power available to the rotor shaft at different wind speeds (Figure 1);

2) Following is the plot of external torque to be provided by motor to achieve the rated rpm (Figure 2). Maximum torque that can be provided by our installed DC Motor is 2.45 N-m. In the Figure 2 green region corresponds to desirable region of operation and red region corresponds to undesirable operational region.

2.3. Experimental Results

1) An experiment with power output is conducted on 700 W SNT-1 micro wind turbine at the premises of

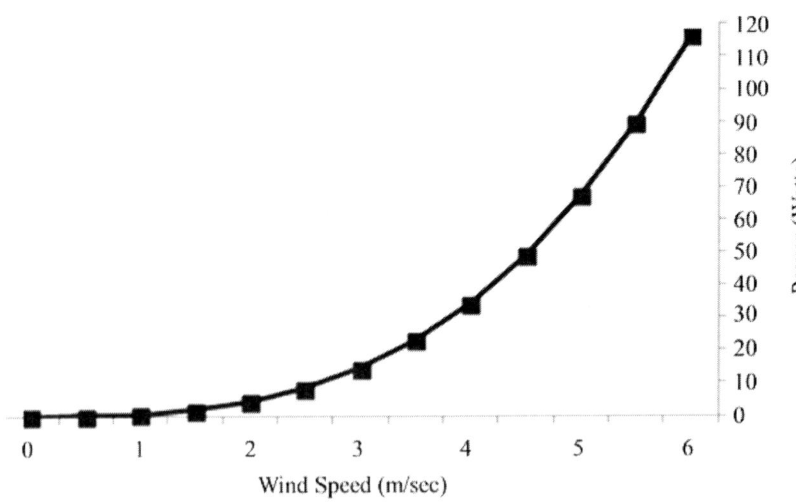

Figure 1. Available rotor power at various wind speeds.

SNT Unit, Udyognagar, Anand, Gujarat and a comparative study of theoretical [11] -[13] and practical results are shown in Figure 3.

2) However in a controlled environment on a test bench of SNT, a 3HP motor is used to rotate the shaft of 400 W generator. The results obtained are contradictory to our theoretical results. We obtained power consumed

is greater than power generated for every value of generator rpm. However, this is not our intention to do, because, the entire energy to rotate the generator shaft is provided by motor in this case. In our method, partial energy would be provided by the wind. Experimental results using external motor without wind are as shown below (Figure 4 & Figure 5).

3) When the above curve of power generated was studied at various equivalent wind speeds (assuming TSR as constant according to the rated value) following results were obtained (Figure 6).

4) Experiment is conducted on Rooftop Wind SNT-1 Wind Turbinein normal mode with low wind speed. Grid Tied Rooftop Windmill installed on the rooftop of GERMI is SNT-1 Wind Turbine with generator Rated Power of 700 W and external Motor Rated Power of 18 W. Results obtained are as follows (Figure 7).

5) In another experiment on SNT-1 Rooftop wind turbine at GERMI, it is observed that power consumed by motor is much higher compared to power generated by the turbine. This is due to the fact that the rotor is purely running on motor and there is no effect of wind on rotor when the motor is in switched on mode. This is due to the current mechanism of spring balance which is being used to align motor with generator axis which prohibits any affect of wind on rotor. Experimental results when the external motor is used are as follows (Figure 8).

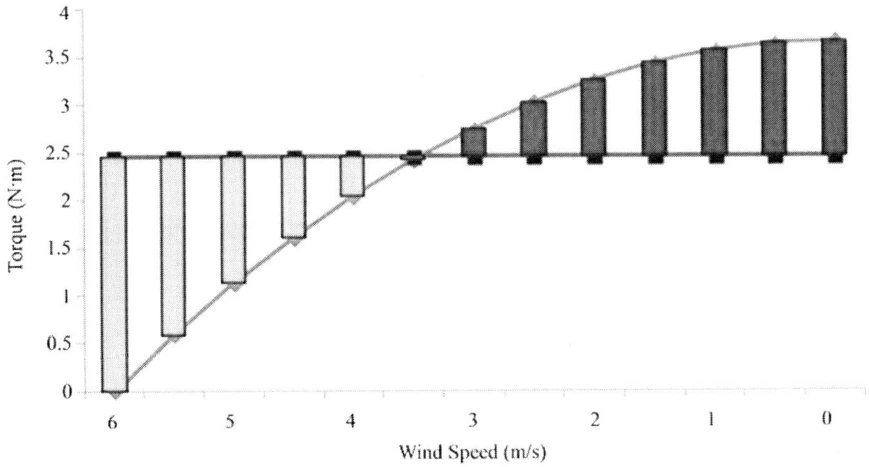

Figure 2. Torque to be provided by external motor at various wind speeds.

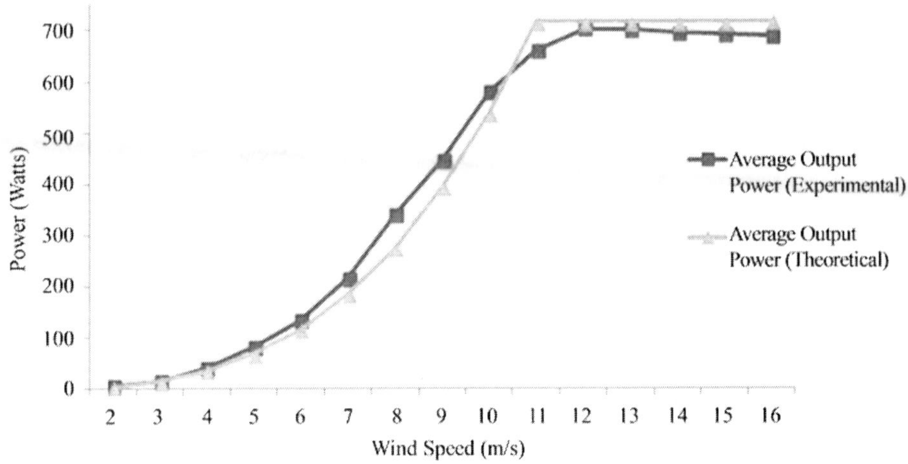

Figure 3. Comparative study of average output power (theoretical & experimental).

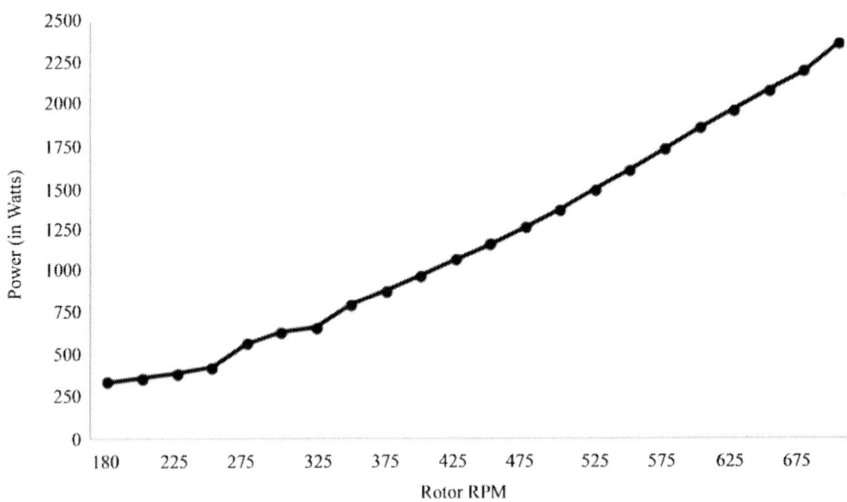

Figure 4. Power consumed by motor at various rotor RPM.

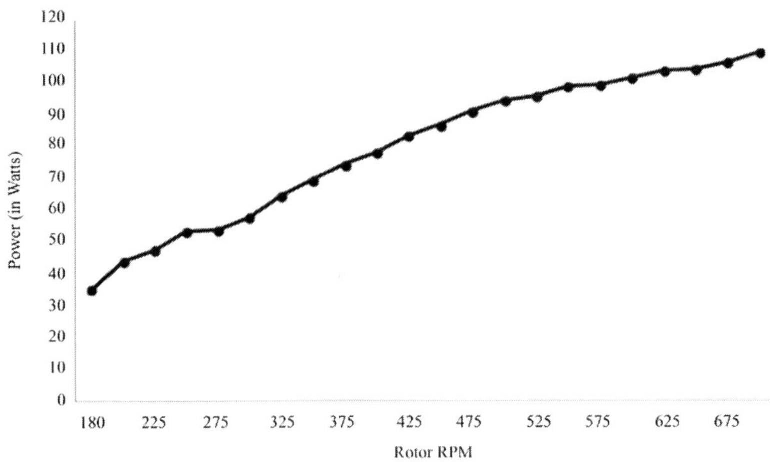

Figure 5. Power generated by generator at various rotor RPM.

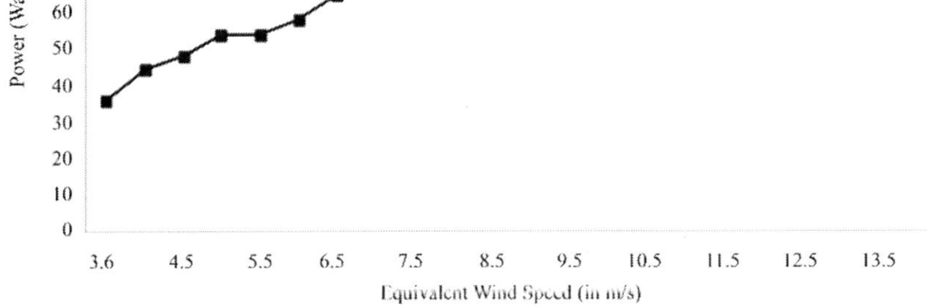

Figure 6. Power generated by generator at various equivalent wind speeds.

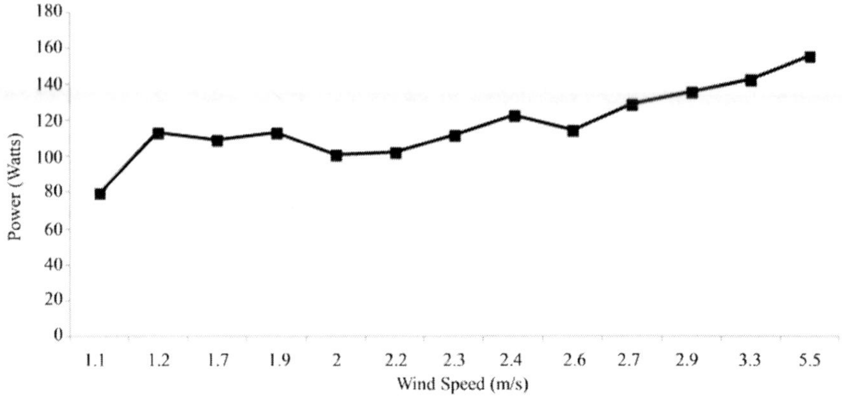

Figure 7. Power generated at various wind speeds (site results).

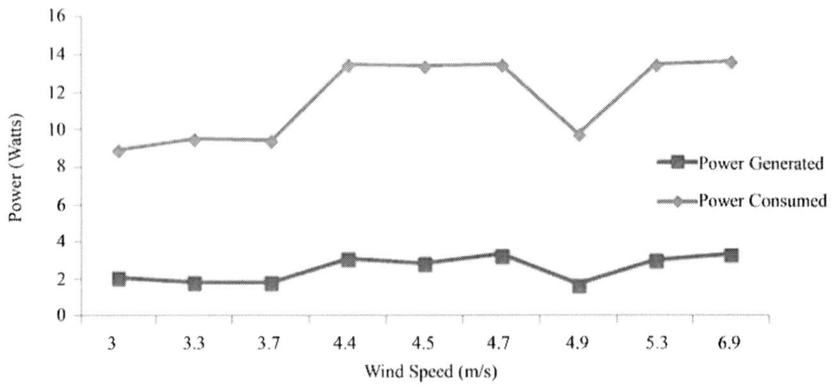

Figure 8. Power generated and consumed at various wind speeds.

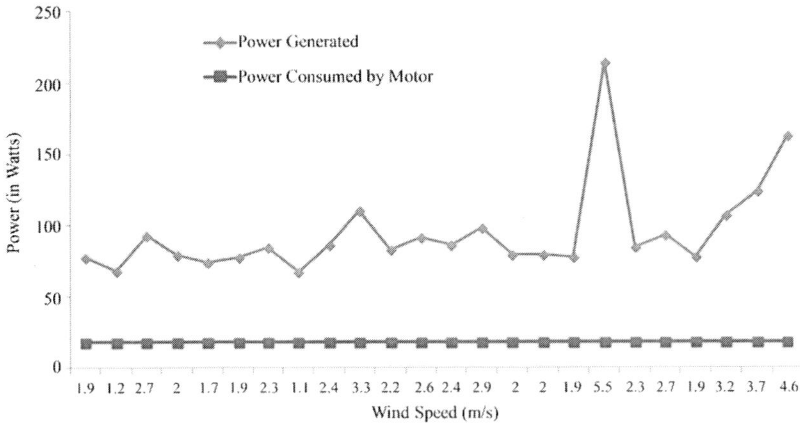

Figure 9. Power generated and consumed at live wind speeds.

Figure 10. 0.7 kW SNT-1 rooftop wind turbine installed at GERMI rooftop.

Figure 11. Grid tied inverter for SNT-1 wind turbine.

Figure 12. Variac for external 12V DC Motor.

2.4. Theoretical Power Gain

The results in Figure 9 are obtained on comparing the results of power generated and power consumed for live wind speed data.

The wind data is recorded for every 2 minutes and the difference between the the two curves actually demonstrate the net power produced by the wind turbine system. The above results clearly show that we gain power in our discussed method and hence our method proves to be fruitful theoretically whereas the experimental feasibility has been partially established but it is yet to be worked upon more extensively.

3. CONCLUSIONS

We have addressed a few issues related to wind turbines and suggested solutions for the same. Hopefully our suggested method will yield into a product that would help changing the wind power scenario at domestic level. SNT-1 wind turbine installed on rooftop of our energy building—GERMI at PDPU campus (Figures 10-12)—is specially designed for low wind speed regime and for the first time attempted and installed a grid tied rooftop micro wind turbine. Further improvement is needed to reach to a level to work on concrete proof. We need to test our suggested method's usefulness to the society as well as to the industry. In this respect, more exhaustive experiments could be devised with data logging facilities. The wind braking and motor system will be upgraded and automated in near future for better results. A control system could also be attached which disconnects the generator to grid when the grid supply is off. During the monsoon season, when the wind velocity is fair enough, our method can be a real asset at least for rooftop wind turbines [14] . The only limitation being it will be commercially beneficial only when the wind speed ranges from 3 to 6 m/s.

ACKNOWLEDGEMENTS

We would like to thank all the GERMI staff for their full cooperation to execute this task. Full support provided by Dr. Sagar Agarvat, Scientist C, Solar Research Wing, GERMI is greatly acknowledged. His coordination and help is very much appreciated without which we could have not completed this work. Thanks are due to Mr. Manyam Telukuntla from UPES Dehradun, a summer intern student at GERMI for assisting in the experimental work. Mr. Ankur Bhatnagar, Mr. Dharmik Lay, Mr. Prashant Gopiyani, Mr. Nilesh Patel and Mrs. Sandhya Radhakrishnan have helped in many ways for successful completion of this project.

REFERENCES

1. Christiner, M., Dobbins, R., Ndegwa, A. and Sivak, J. (2010) Rooftop Wind Turbine Feasibility in Boston, Massachusetts. Worcester Polytechnic Institute, Worcester.

2. Jha, S.N. and Gandhi, J.R. (2006) Low RPM, High Torque Small Wind Generator for Hybrid System. http://www.ibrarian.net /navon/paper/Low_RPM__High_Torque_Small_Wind_Generator_for_Hy.pdf?paperid=7651162

3. Indian Wind Energy and Economy. www.Indianwindpower.com

4. Centre for Wind Energy Technology (C-WET), Chennai. http://www.cwet.tn.nic.in

5. Burton, T., et al. (2001) Wind Energy Handbook. John Wiley & Sons, Ltd., Hoboken.

6. Ahmed, S. (2010) Wind Energy: Theory and Practice, Department of Mechanical Engineering, MNNIT Bhopal, Prentice Hall of India, New Delhi.

7. Johnson, G.L. (2001) Wind Energy Systems. Kansas State University

8. Ragheb, M. (2014) Optimal Rotor Tip Speed Ratio. http://mragheb.com/NPRE%20475%20Wind%20Power%20Systems/Optimal%20Rotor%20Tip%20Speed%20Ratio.pdf

9. McCosker, J. (2012) Design and Optimization of a Small Wind Turbine. Rensselaer Polytechnic Institute, Troy, NY.

10. Ragheb, M. and Ragheb, A.M. (2011) Wind Turbines Theory—The Betz Equation and Optimal Rotor Tip Speed Ratio. Fundamental and Advanced Topics in Wind Power, Dr. Rupp Carriveau (Ed.), InTech. http://www.intechopen.com/books/fundamental-and-advanced-topicsin-wind-power/wind-turbines-theory-the-betz-equation-and-optimal-rotor-tip-speed-ratio

11. Predescu, M., Bejinariu, A., Mitroi, O. and Nedelcu, A. (2009) Influence of the Number of Blades on the Mechanical Power Curve of Wind. International Conference on Renewable Energies and Power Quality (ICREPQ'09), Valencia, 15-17 April 2009, 6p.

12. Sarkar, A. and Behera, D.K. (2012) Wind Turbine Blade Efficiency and Power Calculation with Electrical Analogy. International Journal of Scientific and Research Publications, 2, 1-5.

13. Martinez, J. (2007) Modelling and Control of Wind Turbines. https://workspace.imperial.ac.uk/centreforprocesssystemsengineering/Public/MSc%20Thesis/2007%20MSc%20Thesis/2007.2%20-%20Martinez%20Jasmin.pdf

14. Rivkin, D.A. and Silk, L. (2013) Wind Energy (Art and Science of Wind Power). Jones & Bartlett Learning, Burlington, Mass.

4

CHAPTER

HOLONIC ARCHITECTURE OF THE SMART GRID

Ebisa Negeri[1], Nico Baken[1], Marjan Popov[2]

[1]Network Architectures and Services, Delft University of Technology, Delft, The Netherlands; [2]Electrical Sustainable Energy, Delft University of Technology, Delft, The Netherlands.

ABSTRACT

With the growing concerns about sustainable energy, energy efficiency and energy security, the electrical power system is undergoing major changes. Distributed energy sources are becoming widely available at the lower parts of the grid. As a result, more and more end consumers are transforming from passive consumers to active "prosumers" that can autonomously generate, store, import and/or export power. As prosumers increasingly dominate the power system, the system demands capability that allows enormous number of stakeholders with heterogeneous types to exchange power on the grid. Unfortunately, the classical power system cannot efficiently handle this scenario since it was designed for centralized power distribution. Thus, restructuring the rather old power system is indispensable. In this paper, we apply the holonic approach to structure the smart grid as a system that is bottom-up organized from autonomous prosumers that are recursively clustered at various aggregation layers. Based on this, we present a control architecture of the smart grid using holonic concepts. Our control architecture is characterized by autonomy of the prosumers, distributed control, recursive self-similar control structures at different aggregation levels. Further, we present a service oriented architecture (SOA) framework that models the control functions that make up the holonic control architecture. Our proposed control architecture is tested using a simulation set-up.

Keywords: Smart Grid; Architecture; Holon

1. INTRODUCTION

The electric power system, a vital system supporting the economy of every nation, is undergoing tremendous transformations driven by the growing concerns for increasing the share of green power supply, as well as enhancing the reliability, security and efficiency of the system. As opposed to the classical power system where power is generated at large-scale power plants and transported in a unidirectional way to the distributed consumers, distributed energy sources (DES), such as solar panels and wind turbines, are increasingly penetrating into the lower parts of the power grid. With the availability of the DES, the end customers of the power system are transforming from passive consumers to active "prosumers" that can generate power on site and feed their surplus production back to the grid. According to the European parliament, all new buildings to be built after 2019 will have to produce their own energy on site[1]. Moreover, the foreseen massive presence of electric vehicles (EV) will have a huge impact on the power system. The flexible appliances, energy storage systems, as well as the grid-to-vehicle (G2V) and the vehicle-to-grid (V2G) features of the EVs will offer the prosumers the flexibility to manage the shape of their loads. Given their generation capacity and their flexibility to shape their load profiles, the prosumers tend to autonomously manage their own resources and exchange/ trade power with their surrounding network. As prosumers dominate the system, the power system faces an era of "prosumerization" whereby enormous number and types of stakeholders can autonomously generate, store, import and/or export power. While in the classical power system power is transported in a unidirectional way topdown from the centralized plants to the distributed consumers, in the prosumer-dominated power system power will be primarily exchanged locally between the prosumers in a bidirectional way.

Apparently, restructuring the rather old power system is crucial in order to accommodate the dynamics. Accordingly, an advanced future power grid paradigm is envisaged, which is commonly referred to as the smart grid. The anticipated benefits of the smart grids are listed in the reports of the International Energy Agency (IEA) [1], the National Institute of Standards and Technology (NIST) [2] and the European Smart Grids Technology Platform [3]. Enhancing the reliability, security, efficiency, and resilience of the power system, accommodating all generation and storage options, enabling informed participation of customers, accommodating new products, services and markets are among the benefits. The challenge still remains on how to organize the smart grid in order to realize these benefits.

In this work, we focus on the future prosumer-dominated power system, and study how to organize the smart grids to deliver the expected benefits. We envisage to organize smart grids bottom-up from prosumers. While prosumers autonomously manage their own resources, they can join other prosumers whose profile "complement" their own to exchange power with them. Further, a network of such prosumers might constitute a larger prosumer cluster to gain the power of the collective and bargain with the rest of the grid as a unit. Yet, several clusters of prosumers can form a network at a next aggregation level to form a bigger prosumer cluster and benefit the power of the bigger collective. In this context, it seems that the future power system builds bottom-up through recursive clustering of prosumers at various aggregation levels forming networks of networks. We also propose a suitable control architecture for such a bottom-up organized smart grid. To this end, we embark on concepts from holons and holarchy, which will be described in Section 3.

The contributions of this paper are summarized as follows:

- We employ the holonic concepts to organize the smart grids enabling it to foster seamless accommodation of the foreseen dominance of prosumers.
- We propose a holonic control architecture for our envisaged smart grids.
- We present an SOA framework to support our proposed control architecture and validate it with a simulation set up.

The remainder of this paper is organized as follows. In Sections 2 and 3, we present the related work and the concepts used in our work, respectively. The envisaged smart grid holarchy and the holonic control architecture are described in Sections 4 and 5, respectively. After presenting our simulation scenario and results in Section 5, we provide discussions and concluding remarks in Section 6.

2. RELATED WORK

Recently, various new concepts and developments have been proposed to manage the new trends in the power system. The concept of virtual power plants (VPP) was introduced as a cluster of distributed power sources which are collectively managed [4]. The Microgrid concept [5] was also suggested as a low voltage (LV) distribution system comprising of distributed sources, controlled loads and storage systems that are coordinated to achieve a controllable operation either as an island or connected to the medium voltage (MV) grid.

Provoost et al. [6] introduced the autonomous network (AN), which is a part of the system but its behavior is more or less independent from the rest. AN differs from the microgrid because AN is larger in size and complexity, as well as its aim is primarily optimizing its normal operation. Overbeeke and Roberts [7] proposed the active networks concept, where the network is subdivided into cells that are self managing, but not necessarily self supplying. This concept involves interconnection between the cells that provides more than one power flow paths, and allows rerouting to avoid congestion and to isolate faults. Further, system services are traded along the connections between the cells.

The smart grid concept represents the overall picture of the future power network that is supported by intelligent distributed devices and communication technologies. Tsoukalas and Gao [8] modeled the smart grid as an energy internet comprising of local area grids (LAGs), that are demand based autonomous entities consisting of a convenient mixture of different customers. Huang et al. [9] presented the Future Renewable Electric Energy Delivery and Management (FREEDM) as a power distribution system that interfaces with residential customers and industry customers having distributed renewable energy sources and distributed storage devices. The key technology features of the system are the plug-and-play interface, energy router, and pen-standard based operating system. The proposed system relies on a flat and distributed management architecture.

A market based control concept, named PowerMatcher, is suggested by Kok et al. [10] for supply and demand matching in electricity networks with large penetration of distributed sources. In the PowerMatcher, each device uses its agent to buy or sell power in the electronic market that is implemented in a distributed manner using a tree structure. On the other hand, service oriented architecture (SOA) is gaining attention in power system control. Vaccaro et al. [11] have proposed a web-servicebased framework for integrated microgrid (MG) control, modeling, and monitoring. At the core of their framework is the MG engine which executes the MG management in a geographically distributed scenario.

Although their scale, complexity and intelligence varies, all the above concepts and technologies have their own contributions in managing the distributed sources. One concept could be more applicable in some settings than the other, and vice versa. Moreover, some of the concepts could be combined since they complement each other. The smart grid could be an umbrella to appropriately combine the concepts, thereby achieve a heterogeneous and intelligent power system that delivers its expectations. For instance, the FRIENDS, VPP, microgrid, AN, and active networks could form sub-networks of the smart grid, while FREEDM provides the power delivery infrastructure supporting the sub-networks and PowerMatcher delivers the transaction services.

In this work, we study how the smart grids can be organized to accommodate the foreseeable dominance of prosumers. We endeavor to propose an architecture of the smart grid that tends to converge the various proposed concepts. We model the heterogeneous types of components of the smart grid as prosumers. Based on this, we propose to organize the smart grid bottom-up from prosumers, by recursively clustering them at various aggregation layers. Moreover, we also propose a control architecture that fits our proposed organization of smart grids. Our prosumer model and control architecture are built based on the concept of holons and holarchy that is quite suitable to capture the properties of the future smart grids.

3. RELATED CONCEPTS

3.1. Prosumer

In this work, prosumer is a general term that refers to a system that autonomously manages its resources and is capable of bidirectional exchange of power with its surrounding network. A prosumer could be as simple as a household that autonomously manages its resources (such as appliances, distributed sources, storage systems, electric vehicles, etc.). Other systems such as a virtual power plant, a microgrid, an energy community, an autonomous network, a power plant, etc. can also be regarded as prosumers. A prosumer can comprise sub-parts that are themselves prosumers, while at the same time can become a part of a larger prosumer. With the aid of the intelligent information and communication technology (ICT) capabilities of the smart grid, a prosumer can orchestrate its resources as well as coordinate itself with its surunding network to achieve desirable internal and system wide goals.

3.2. Holon and Holarchy

The word holon was coined by Koestler [12], and it refers to a logical entity that is both a whole and a part. For example, a cell in your body is a holon because it is a disnct living entity that has a distinct cell wall defining its interface with the rest of the world. However, a cell is composed of smaller holons such as nucleus and chromosome that are also separate entities. Yet, a group of cell holons together form a higher aggregation layer tissue holon, a group of tissue holons form an organ holon, and so on.

A holon is basically autonomous, cooperative and recursive. Holons exhibit self-similar recursive structure and are organized in various aggregation layers to form a holarchy as shown in Figure 1. Holarchies are recursive in the sense that a holon in a holarchy can in itself be an entire holarchy of other holons. In a holarchy, each holon can function autonomously, thereby enabling formation of a distributed system. While functioning autonomously, a holon can cooperate with other holons in the holarchy to achieve mutual goals. In a dynamic environment, holons can also be dynamically reorganized, thereby making the holarchy more robust to changes adapting itself to its environment. These properties of the holonic concept make it quite appealing to organize very complex systems. The holonic approach has been proposed to organize various systems, some of them are organizational modeling, manufacturing, sensor networks, traffic light management, etc. ([13-17]).

3.3. Service Oriented Architecture (SOA)

SOA is an architecture that models an application as a set of loosely coupled, flexible, reusable and adaptable functional units called services, whereby the services are interrelated through well-defined interfaces and contracts. The main advantages of SOA include platform-independence of the interfaces that aids interoperability, the flexibility and reusability of the services, and its easy adaptation to evolutionary changes in the application. While SOA can be implemented using different technologies, the web services are the most popular one because of wide acceptance of its standards. Web service technology exposes web services, which are sets of operations on the web; thereby allowing high degree of integration among them. These features of SOA and web services technologies are attractive to model the control functions of the holonic control, as will be discussed later.

4. THE HOLONIC SMART GRID

While the smart grid has diverse types of components, such as households, microgrids, energy communities, etc., each of the components can be seen as a prosumer. The prosumers tend to autonomously manage their resources. Moreover, they tend to aggregate together and form a bigger cluster to facilitate local power exchange as well as to gain a larger power of the collective to trade power with their surrounding. For instance, a group of prosumer households can cluster together to form a neighborhood energy community, which is a larger prosumer. Likewise, a group of neighborhood energy communities can form a district energy community (yet a larger prosumer) at the next aggregation level. This clustering could further be recursively repeated at various aggregation levels. Moreover, the prosumers can dynamically reorganize to efficiently adapt to the change in their environment. Apparently, the autonomy, aggregation into layers, recursive property, and dynamic adaptation of the prosumer based smart grids closely matches the properties of a holonic system. Accordingly, we model each prosumer as a holon and the entire smart grid as a holarchy. The smart grid holarchy has the following major features that contribute to the overall efficiency of the system.

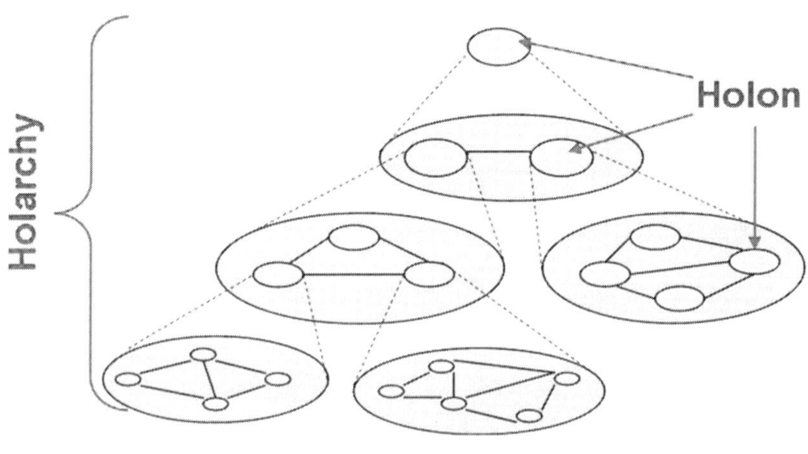

Figure 1. A holarchy—organization of holons.

4.1. Autonomy/Self Management

As prosumers own production capacity and acquire the flexibility to manage their own load profiles, they tend to autonomously manage their resources to optimize their interest. This is in line with the growing interest towards active participation of customers. A prosumer holon might also choose to be part of a bigger prosumer holon and exchange power with its surrounding, or operate as a self-supplying islanded unit. The autonomy of the prosumer holons aids a distributed control capability of the system that would be otherwise very difficult to control centrally. Autonomy might also increase the consciousness of the prosumer holon, thereby making it more cautious about its consumption. Autonomy of the prosumers might make coordination not easy to achieve, but this can be achieved using machine learning [18], game theory [19] and mechanism design [20] based approaches.

4.2. Recursive Aggregation

As mentioned earlier, aggregation is important for prosumers to facilitate local power exchange between themselves, and to gain a bigger bargaining power with the rest of the grid. Accordingly, the prosumer holons at a lower aggregation level may be self-organized to constitute a bigger prosumer holon. For the same reasons, the newly formed prosumer holon can still be connected to its peers in a network in the next higher aggregation layer to form yet a bigger prosumer holon. Further bottom-up grouping of such holons in networks-of-networks continues recursively in higher aggregation levels and eventually constitute the overall holarchy of the smart grid (Figure 2). Simon [21] has shown that this recursive clustering at different aggregation layers is inherent behavior of all complex systems. We refer to the smaller holons that form a bigger holon as the sub-holons of the bigger holon, whereas, the bigger holon on the next aggregation layer that contains a holon is referred to as its super-holon.

The bottom-up organization at different aggregation layers provides efficient structure that encapsulates and simplifies the coordination of the system, making it scalable. Moreover, the aggregation of prosumer holons into super-holons might increase the reliability of the collective as the profiles of the individual holons might complement each other. Further, the local power exchanges between the prosumer holons reduce both the transportation losses and the costly investment costs to upgrade the transmission capacity to meet the ever increasing demands.

4.3. Dynamic Reconfiguration

Over an interval of time, a prosumer holon may be in the state of either self sufficiency, surplus production, or excess demands. Thus, the prosumer holons can leave one super-holon to join another one over a period of time to look for a set of holons that complement their profiles, thereby optimizing their resource utilization. Erol-kantarci et al. [22] have shown that a set of microgrids could be dynamically reorganized during a day into clusters according to their profiles to optimize the utilization of their renewable resources. Moreover, prosumer holons can join a super-holon for some time intervals during a day and operate in islanded mode at other intervals, for example in response to potential faults. Such flexibilities lead to a dynamic reconfiguration of the smart grid holarchy over a period of time. The dynamic reconfiguration contributes to the self-healing, dynamic recovery, and increase efficiency of the prosumer holon.

4.4. Intelligence and Communication

A prosumer holon needs to coordinate its components, as well as coordinate itself with its surrounding to optimize utilization of resources and to ensure system stability.

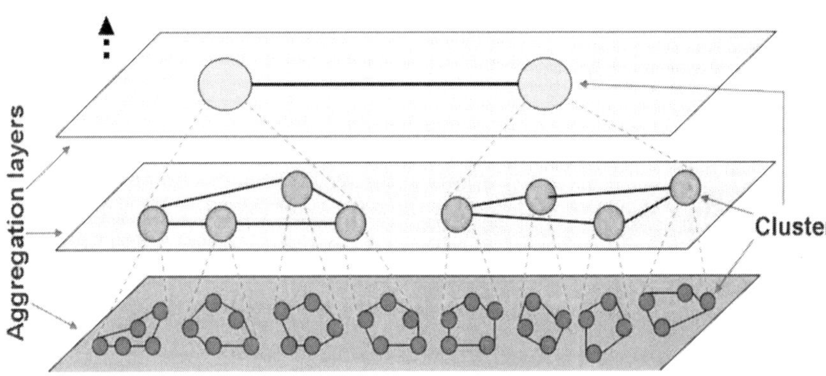

Figure 2. Bottom-up organization of the prosumers to form the smart grid holarchy.

This coordination involves large information flows and requires intelligent information processing units that constitute the "nerve system" of the "organic" power system. Accordingly, each prosumer holon requires communication capabilities and intelligence to make decisions based on the gathered information. Indeed, an appropriate communication infrastructure needs to be in place. The IEC- 61850 standard [23] is developed for communication in substations. A review of the types of intelligence required as well as the candidate communication technologies for smart grids is presented in [24].

4.5. Security and Privacy

Cyber security is regarded as one of the biggest challenges in smart grids [2,3]. The enormous exchanges of information and control messages raise security concerns for the prosumers. Thus, the information and control message exchanges need to be protected against unauthorized access and guarded from malicious hackers. Therefore, the system requires a secure communication protocol. Moreover, privacy is becoming a major concern as the data exchanged might reveal personal information such as habits, activities, and preferences of the individual. For instance, the study in [25] has revealed that personal information can be estimated with high accuracy based on the meter reading of a household, even with relatively unsophisticated hardware and algorithms. Several solutions have been proposed to overcome security and privacy issues in smart grids, and their review is presented in [24]. The IEC61850 standard also supports secure communication in substations.

5. THE HOLONIC CONTROL ARCHITECTURE

Given the nature of the envisioned smart grid, its control architecture needs to exhibit the following properties: 1) sufficient autonomy of each prosumer holon to manage its own resources; 2) recursive structure where a prosumer is composed of smaller prosumers at the lower aggregation layer, and at the same time be part of a larger prosumer holon at a higher aggregation layer; 3) layered structure to facilitate coordination between different aggregation layers; 4) dynamic reconfiguration to runtime topology change. The holonic architecture, as explained before, nicely fits the envisioned system because it possesses all of the properties required by the control architecture.

Our holonic control architecture maintains the proposed structure in the envisioned power system. In the control domain, a control-holon represents the control system of a prosumer holon. Thus, holonic control units will be recursively organized in a bottom-up structure to form a complete control holarchy of the smart grid. When the prosumer holon has prosumer sub-holons, its controlholon is also composed of the corresponding control-subholons (i.e. control systems of sub-holons).

In addition to controlling the internal dynamics of its prosumer holon, a control-holon needs to interact with the control-holons of its peer prosumer holons that are in the same super-holon as well as with that of its super-holon. Moreover, a control-holon needs to interact with the network operator operating at its aggregation level, regarding the power quality, infrastructure constraints, etc. Thus, the control functions of the control-holons are basically the same independent of their aggregation levels. This reveals the self-similarity (recursive property) of the control holarchy. Figure 3 shows the functional architecture of a single control-holon. Corresponding to each of the control functions in the control architecture shown in Figure 3, we define a corresponding web service in the SOA model together with its possible operations. The control functions and their corresponding web services are described subsequently.

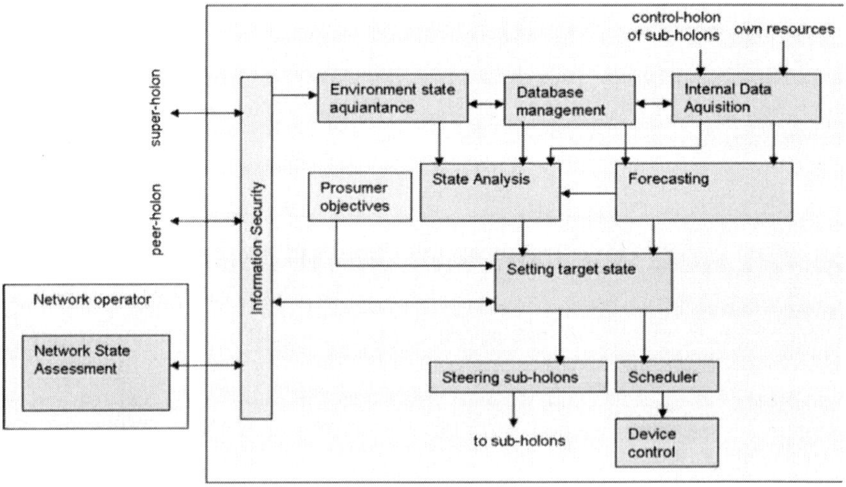

Figure 3. Architecture of a control holon.

5.1. Environmental State Acquaintance

A prosumer holon needs to be well acquainted with its environment in which it is operating to make optimal decisions. Acquiring information about the load profiles of its peer prosumers, market prices, incentives, etc. is helpful to plan optimal transactions with its environment. When a super-holon sets a target state (of, for instance, the aggregate load profile), it might generate control signals, incentives, etc. for its sub-holons, based on their mutual agreements, to achieve the desired goal. Therefore, this control function is responsible for acquiring such information and presents it to the relevant control functions in a useful shape. The web service corresponding to this control function is EnvironmentSA-srv. To provide its functionality, the web service involves several operations. Some of the operations that compose the EnvironmentSA-srv web service are GetLoadProfile-opr, to acquire the load profile in its environment, and GetMarketState-opr, to acquire the market price in its environment.

5.2. Internal State Acquisition

A prosumer holon needs to have up-to-date information about its own constituents in order to make optimal decisions. This function handles acquiring internal data about the parameters of the resources of the prosumer (such as distributed generations, storage systems, etc.), their current available capacities, the state (such as, the load profiles, surplus productions, etc.) of the prosumer sub-holons, etc. Corresponding to this control function, we define the InternalDA-srv web service. Below are some of the operations constituting the web service:

- GetResourcePar-opr: to get the parameters of a resource, such as the storage capacity of a battery.
- GetResourceState-opr: to acquire the current state of a resource, e.g. the current charge level of a battery.
- GetSubProfile-opr: to get the load profile of a prosumer sub-holon.

5.3. Database Management

This function is responsible for storing, retrieving and updating the relevant internal and external data. The web service that provides this function is DatabaseMgt-srv. Accordingly, the following operations can be identified as some of its components: RdData-opr, RtData-opr, UpData-opr, and DlDalata-opr, that are used to read, write, update, and delete data, respectively.

5.4. Forecasting

This control function takes care of forecasting both the internal profiles such as generation profile of its local power sources, its local power demands, the load profile of its sub-holons (if any), as well as the load profile of its environment. The equivalent web service is Forecast-srv, which could compose the operations ForecastProduction-opr, ForecastDemand-opr and ForecastEnv-opr that are responsible for forecasting the local production, local demand, and the load profile of its environment, respectively.

5.5. State Analysis

Based on the inputs from the environment state acquaintance, the internal data acquisition, and the forecasting control functions, this control function analyzes the resulting state of the system. This allows the control holon to assess the possible load profiles, the flexibility in the load profiles, the synergetic potential of the various components of the resources and the load profiles, etc. We define StateAnalysis-srv as the web service that takes care of this control function. The following are some of the operations that could compose this service:

- AssFlexibility-opr: to assess the flexibility of the load profiles.
- AssSynergy-opr: to assess the synergetic potential of the resources.
- CompState-opr: to compute possible states of the aggregate load profiles of the holon.

5.6. Setting Target State

After acquiring the state estimation information, this function sets the target state for the prosumer. The target state specifies the target load profiles, amount of power to be imported/exported, whether to operate connected or as an island (where applicable), etc. To set the target set, the control function uses as input, among others, the objectives of the prosumer, which could be cost minimization, increasing comfort, increase composition of green energy, etc. The web service that handles this function is TargetState-srv, and some of the operations that compose it are SetTargetProduction-opr and SetTargetDemand-opr that take care of setting the target production and demand profiles, respectively, of the prosumer holon.

5.7. Steering Sub-Holons

Based on the target state proposed, this control function generates relevant signals for the sub-holons that would help to achieve the goal. The signals generated could be incentives or control signals depending on the type of the terms and conditions in the contract of the prosumer with its sub-holons. Corresponding to this control function, we define a web service called SteerSubHolon-srv. The operations that constitute the service could be GenControlopr and GenIncentive-opr to generate downstream control signals and incentive signals, respectively, for its sub-holons.

5.8. Scheduler

The scheduler determines the optimal load profiles of the schedulable resources of the prosumer to achieve the target state. Its equivalent web service is Scheduler-srv. An operation that could make up the web service is GenSchedule-opr, which is used to generate the optimal schedule of the resources.

5.9. Device Control

The device control function translates the schedule of the resources into the actual control signals of the devices that are scheduled. The DeviceControl-srv web service takes care of this, while the operation SetDeviceControlopr could be employed to set the device controls accordingly.

5.10. Information Security

The control function maintains the security of the information and control message exchanges of the controlholon with its environment. Its corresponding web service is Security-srv, which could have the following operations.

- EncryptInfo-opr: to encrypt the information to be sent.
- DecryptInfo-opr: to decrypt the received information.
- CheckAuthorization-opr: to check the authorization of the sender of the information.

5.11. Network State Assessment

This control function takes care of assessing the state of the power network over which the prosumer holons are interconnected. This includes analyzing the voltage levels at various nodes in the network, the loading of the infrastructures, etc. It also deals with contingency analysis, alarms about faults and potential risks in the environment. This functionality is provided by the network operator through the web service NtkStateAssessment-srv, which could be composed of operations such as, AssessNetk-opr, to assess the state of the network, and SdAlarms-opr, to send alarms to the corresponding holons.

6. SIMULATION SETUP

We considered a system where our proposed control architecture can be applied. Our experimental setting consists of prosumer holons that are aggregated into three layers. At the lowest level are a set of prosumer households (PH) that form a neighborhood energy community at the next aggregation level. The neighborhood energy community (NEC) is in turn part of a district energy community (DEC) at the next higher aggregation layer. Although their aggregation levels differ, the PHs, the NEC, and the DEC are all autonomous prosumer holons. The control-holon of the PH autonomously manages its resources to achieve a desired load profile of the PH. The control-holon of the NEC coordinates the control-holons of its PHs, by providing incentives, to achieve a desirable aggregate load profile that maximizes the benefit of the NEC in response to the incentive received from the control-holon of DEC.

The power flow in the power grid is managed by a network operator (NO). The NO interacts with the relevant control holons to make sure that the power flows in the network do not violate the infrastructure constraints. Further, it generates appropriate control signals for the relevant control-holon whenever a fault occurs in the network.

The sketch of the interconnection between the described holons is depicted in Figure 4. The figure shows both the physical grid connection (solid lines) and the corresponding network of the control-holons (doted lines). The control-holons (CH) in the figure could be implemented in intelligent devices that are capable of both making intelligent decisions and communicating across a communication infrastructure. For example, the CH of a prosumer household can be implemented in the home energy management system [26]. In our experimental setting, the control-holons are hosted by computers that are interconnected by IP network.

A Case Example

We present a case example that is implemented in the setting described above. Load management aiming at peak load reduction in a NEC that is composed of PHs is considered. Each of the PHs owns a solar panel, a micro-CHP unit, an electric vehicle and other electricity consuming appliances. The DEC wants to minimize its aggregate peak load and uses a control signal to steer each of its constituent NECs to reduce its peak load. The NEC

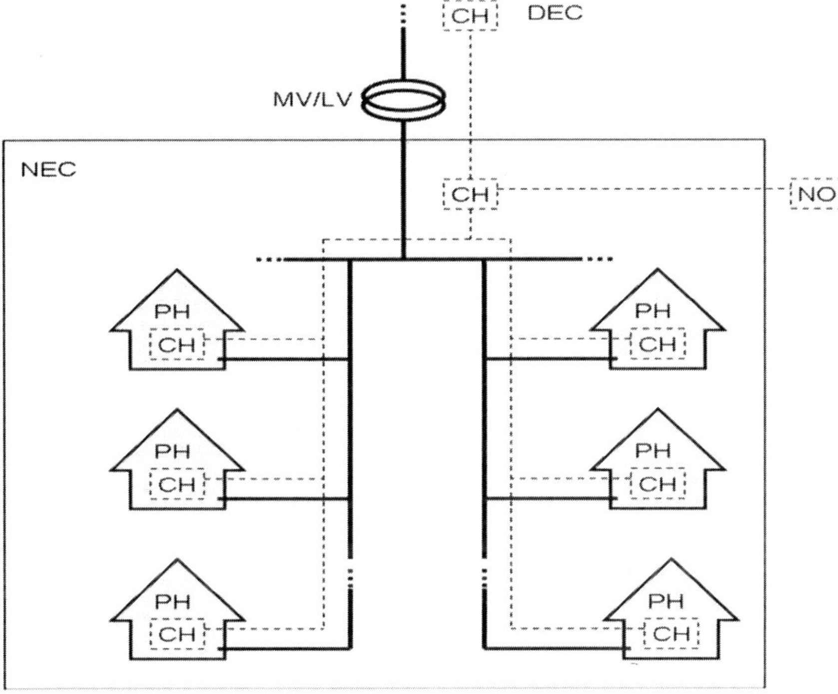

Figure 4. A sketch of the simulation setup (CH = control holon).

provides incentives to the PHs to yield peak reduction. For this purpose, we use the dynamic pricing based incentive mechanism that was proposed in [27]. The PHs tends to minimize their cost. Hence, each PH schedules the (dis)charging of its electric vehicles to minimize its cost in response to the pricing incentive, and then sends its resulting load profile back to the NEC. After receiving the response of all the PHs, the NEC modifies the pricing incentive based on the aggregate load profile and sends it to again to the PHs. This procedure is repeated for a fixed number of iterations. Eventually, the NEC chooses the price vector that yields the lowest peak load, and the corresponding load profiles are adopted. The control agent of the NO employs a low voltage power grid simulator tool called Gaia [28], which is commonly used by network operators in the Netherlands, such as Alliander, to assess the state of the network. For our simulations, we obtained data representing field measurements of the power demand of the appliances as well as the power production of the solar panels and the micro-CHPs from Alliander[2]. Each of these measurements is taken once every 15 minutes over a duration of one day (24 hours). Following this, a time step in our

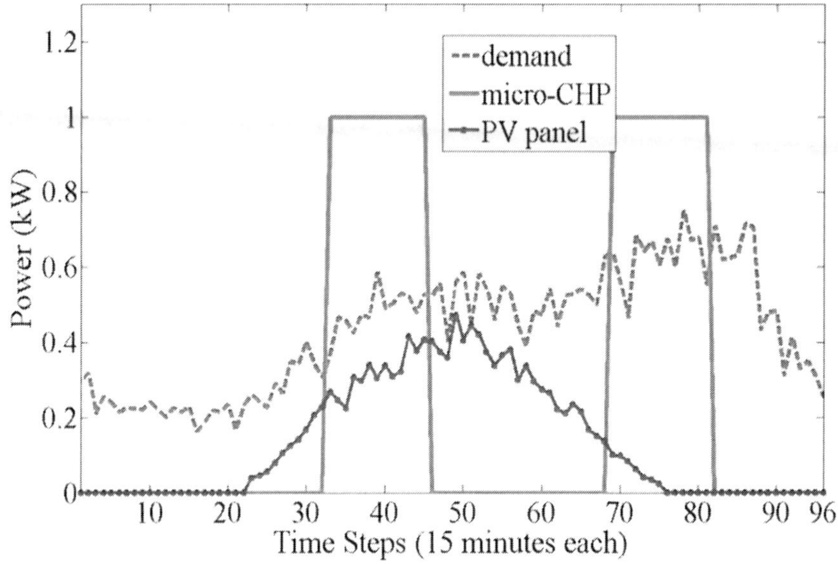

Figure 5. A sample prosumer household demand and production profiles.

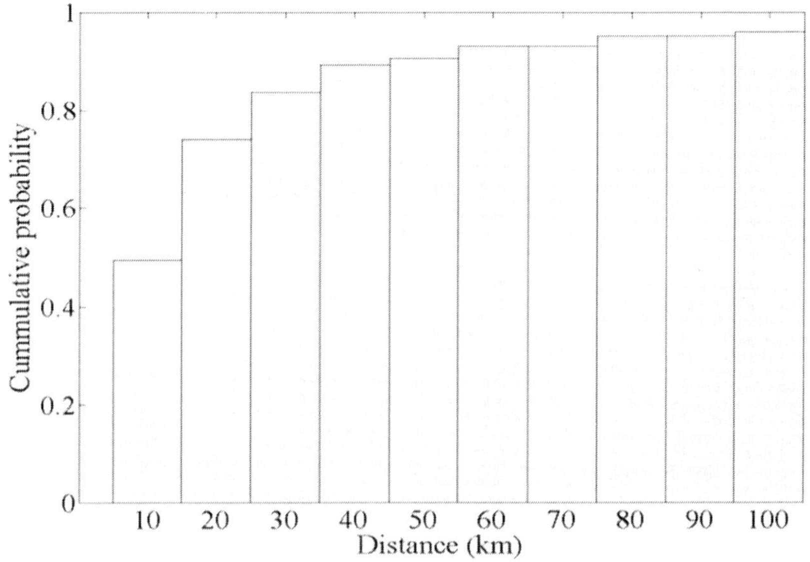

Figure 6. Distribution of the distance traveled per day per car driver.

simulation refers to a 15 minutes time interval, and we have 96 time steps during 24 hours. While the patterns of the daily commute distance, the arrival at and departure from home of the EV drivers are derived from the 2009 mobility research of the Netherlands [29], the parameter specifications of the EVs are adopted from [27]. The plots of some of these data are depicted in Figures 5 and 6. Using the proposed control architecture, the following control steps could be involved to achieve the peak minimization in the aggregate load profile of the NEC.

1) The control-holon (CH) of the DEC, the superholon of the NEC, generates and sends a control signal to the CH of the NEC to reduce its peak load. The CH of the DEC uses the SteerSubHolon-srv web service to do so.
2) The CH of the NEC receives the signal from its super-holon using the EnvironmentSA-srv web service, and sets a target state that reduces peak load accordingly, using the TargetState-srv.
3) The CH of the NEC receives the load profile of its sub-holons (PHs) using the internalDA-srv web service.
4) The CH of the NEC uses the SteerSubHolon-srv web service to generate a price vector based on the aggregate load profile of its sub-holons, and sends it to the CHs of the PHs.
5) The CH of each PH receives the price vector from its super-holon using the EnvironmentSA-srv web service, and sets target state that minimizes its cost based on the received price vector, using the TargetState-srv web service.
6) The CH of each PH receives the data about the capacity and state of its resources using the InternalDA-srv web service. It then sets its target state, using the TargetStatesrv web service to minimize its cost based on the received price vector, and subsequently schedules, using the Scheduler-srv web service, the (dis)charging of its electric vehicle accordingly. Next, it sends the resulting net load profile of the PH back to the CH of the NEC.
7) Steps 3 - 6 represent the negotiation that takes place between the NEC and the PHs to arrive at a situation that yield mutual benefits for the parties. These steps are repeated for a fixed number of iterations. Eventually, the CH of the NEC selects the price vector that yielded the lowest aggregate peak load, together with the corresponding load profiles.

8) The CH of the NEC requests the NO to analyze the network state under the resulting load profile.

9) The NO employs the NtkStateAssessment-srv web service to analyze the resulting network state by using Gaia tool. Afterwards, it sends the result of the analysis back to the CH of the NEC.

10) If the NEC receives a "feasible state" network state report, then the resulting load profile is reported to the CH of the DEC and the selected price vector is sent to the PHs as a final price vector. Subsequently, the CHs of the PHs use the Device Control-srv web services to control their electric vehicle (dis)charging according to the schedule corresponding to the final price vector.

11) If the received network state is "infeasible state", then the CH of the NEC will check the network state with the load profiles resulting from the other price vectors until it receives a "feasible state" reply. If none of the load profile yields a healthy network state, then it will generate and send a control signal to the PHs to reduce peak load.

After implementing the web services using the Java Web Services, the above simulation is executed with a NEC consisting of five PHs, and the resulting aggregate load profile is depicted in Figure 7. As can be observed form the figure, coordination of the NEC using our holonic control architecture has yielded a significant reduction (as large as 54%) in aggregate peak load compared to the uncoordinated scenario where the PHs charge their EVs as soon as they return home from their previous commute. While in our case the Steer Sub Holon-srv service uses the dynamic pricing model proposed in [27] to provide the incentives for the PHs in the NEC, other incentive mechanisms can also be used to achieve a desired aggregate load profile.

7. DISCUSSIONS AND CONCLUSIONS

The trends of transition in the electric power system are leading to an era of prosumerization whereby myriad of stakeholders autonomously generate and exchange power with each other on the power grid. While the classical power system cannot handle the dynamics since it was designed for centralized generation and unidirectional topdown power flows, the future smart grids are expected to efficiently manage the foreseeable era of prosumerization. To this end, we have proposed to organize the smart grids based on the concepts of holons and holarchy that effectively capture the features of the transitions.

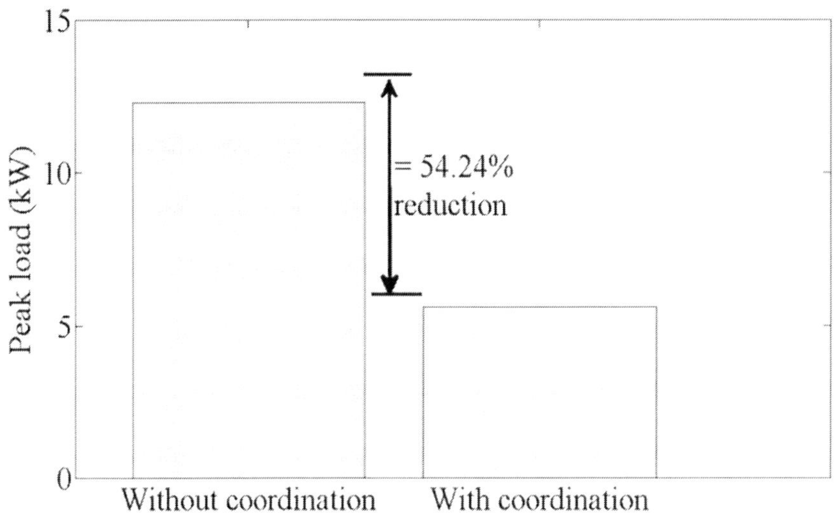

Figure 7. The peak load of a NEC under two scenarios: without coordination and with coordination employing our holonic control architecture.

The major features of our holonic smart grids include sufficient autonomy of each prosumer to manage its own resources; the recursive structure where a prosumer is composed of smaller prosumers at the lower aggregation layer, and at the same time be part of a larger prosumer at a higher aggregation layer; dynamic reconfiguration to runtime topology change; intelligence and communication capability of prosumers, and maintaining security and privacy of prosumers. The prosumer holons are recursively aggregated into clusters to eventually constitute the smart grid holarchy. Corresponding to the envisaged holonic smart grid, we have proposed a holonic control architecture that conveniently fits the aforementioned features. The control architecture is composed of control-holons corresponding to the prosumer holons that are recursively organized to yield control-holarchy of the smart grid. After specifying relevant control functions of our control-holon, we have defined the corresponding web services based on Service Oriented Architecture (SOA) model. Eventually, the SOA model was implemented using Java web services and tested in a simulation setup yielding interesting results.

In order to implement our proposed smart grid holarchy, there are technical, marketing and regulatory requirements that need to be met. The requirements are in line with the trends in the smart grid developments. One of the requirements is the intelligence and communication capabilities of the prosumers. The current developments in smart meters promise availability of intelligent units that are capable of two way communication [30]. Various developments exist with regards to communication technologies that can be deployed in smart grids, comparisons of the candidate communication technologies are presented in [24]. Another requirement is supporting the bidirectional power flows in the grid. The Future Renewable Electric Energy Delivery and Management (FREEDM) [9] is among the promising developments that address this issue. The capability of dynamic reconfiguration is another requirement. The work presented in [22], which allows a dynamic reconfiguration of a set of microgrids is an interesting development in this regards. With the myriad of prosumers trading electrical power, the challenge lies in how to coordinate the market. The PowerMatcher technology proposed in [10] seems a suitable candidate to adopt for this purpose. Moreover, several solutions have been proposed that can be employed to overcome the security and privacy concerns of the prosumers [24]. Another important requirement is the presence of a suitable regulation that allows the autonomous prosumers to trade power on the grid. Given the trends in the deregulation of electricity markets over the recent past years and the ongoing efforts, although the extents of deregulation vary from one country to another, one can be optimistic to foresee a conducive scenario.

In general, our proposed smart grid holarchy is a suitable architecture to accommodate the foreseeable era of prosumerization. Moreover, the existing developments and the ongoing efforts in various aspects of the smart grids are in line with the requirements for its implementation.

REFERENCES

1. International Energy Agency, "Technology Roadmaps Smart Grids," 2011. http://www.iea.org/publications/freepublications /publication /smartgrids_roadmap.pdf

2. National Institute of Standards and Technology, "NIST Framework and Roadmap for Smart Grid Interoperability Standards, Release 1.0," 2010. http://www.nist.gov/public_affairs/releases/upload/smartgrid_interoperability_final.pdf

3. European Smart Grids Technology Platform, "Vision and Strategy for Europe's Electricity Networks of the Future," 2006. ftp://ftp.cordis.europa.eu/pub/fp7/energy/docs/smartgrids_en.pdf

4. K. Dielmann and A. van der Velden, "Virtual Power Plants (VPP)—A New Perspective for Energy Generation?" Proceedings of the 9th International Scientific and Practical Conference on Modern Techniques and Technologies, 2003, pp. 18-20.
5. N. Hatziargyriou, N. Jenkins, G. Strbac, J. A. P. Lopes, J. Ruela and A. Engler, "MICROGRIDS—Large Scale Integration of Micro-Generation to Low Voltage Grids," CIGRE C6-309, 2006.
6. F. Provoost, J. Myrzik and W. Kling, "Setting Up Autonomous Controlled Networks," 39th International Universities Power Engineering Conference (UPEC), Vol. 3, 2004, pp. 1190-1194.
7. F. Overbeeke and V. Roberts, "Active Networks as Facilitators for Embedded Generation," Cogeneration and On-Site Power production, Vol. 3, No. 2, 2002, pp. 37- 42.
8. L. Tsoukalas and R. Gao, "From Smart Grids to an Energy Internet: Assumptions, Architectures and Requirements," 3rd International Conference on Electric Utility Deregulation and Restructuring and Power Technologies, Nanjing, May 2008, pp. 94-98.
9. A. Huang, M. Crow, G. Heydt, J. Zheng and S. Dale, "The Future Renewable Electric Energy Delivery and Management (FREEDM) System: The Energy Internet," Proceedings of the IEEE, Vol. 91, 2011, pp. 133-148.
10. J. Kok, C. Warmer and I. Kamphuis, "Power Matcher: Multiagent Control in the Electricity Infrastructure," Proceedings of the 4th International Joint Conference on Autonomous Agents and Multiagent Systems, Utrecht, 25- 29 July 2005, pp. 75-82.
11. A. Vaccaro, M. Popov, D. Villacci and V. Terzija, "An Integrated Framework for Smart Microgrids Modeling, Monitoring, Control, Communication, and Verification," Proceedings of the IEEE, Vol. 99, 2011, pp. 119-132.
12. A. Koestler, "The Ghost in the Machine," Penguin Group, New York, 1990.
13. G. Zhang and W. Li, "Flexible Holonic Organization Modeling and Cultural Evolution," 4th IEEE International Conference on Wireless Communications, Networking and Mobile Computing, Dalian, 12-14 October 2008, pp. 1-5.
14. H. Van Brussel, J. Wyns, P. Valckenaers, L. Bongaerts and P. Peeters, "Reference Architecture for Holonic Manufacturing Systems: PROSA," Computers in Industry, Vol. 37, No. 3, 1998, pp. 255-274.

15. M. Moghadam and N. Mozayani, "A Street Lighting Control System Based on Holonic Structures and Traffic System," IEEE 3rd Inernational Conference on Computer Research and Development (ICCRD), Shanghai, 11-13 March 2011, pp. 92-96.

16. Y. Ye, V. Hilaire, A. Koukam and C. Wandong, "A Holonic Model in Wireless Sensor Networks," IEEE International Conference on Intelligent Information Hiding and Multimedia Signal Processing, Harbin, 15-17 August 2008, pp. 491-495.

17. N. Baken, "Holons and Holarchies," TEDx Amsterdam 2009. http://www.tedxamsterdam.com/2009/video-nico-bakenon-holons-and-holarchies/

18. D. O'Neill, M. Levorato, A. Goldsmith and U. Mitra, "Residential Demand Response Using Reinforcement Learning," 1st IEEE International Conference on Smart Grid Communications, Gaithersburg, October 2010, pp. 409-414.

19. C. Ibars, M. Navarro and L. Giupponi, "Distributed Demand Management in Smart Grid with a Congestion Game," 1st IEEE International Conference on Smart Grid Communications, Gaithersburg, 4-6 October 2010, pp. 495-500.

20. E. Gerding, V. Robu, S. Stein, D. Parkes, A. Rogers and N. Jennings, "Online Mechanism Design for Electric Vehicle Charging," 10th International Joint Conference on Autonomous Agents and Multi-Agent Systems, Taipei, 2-8 May 2011, pp. 811-818.

21. H. Simon, "Architecture of Complex Systems," Proceedings of the American Philosophical Society, Vol. 106, No. 6, 1962, pp. 467-482.

22. M. Erol-kantarci, B. Kantarci and H. Mouftah, "Reliable Overlay Topology Design for the Smart Microgrid Network," IEEE Transactions on Networks, Vol. 25, No. 5, 2011, pp. 38-43.

23. Alstom Grid, "Network Protection and Automation Guide," 2011.

24. X. Fang, S. Misra, G. Xue and D. Yang, "Smart Grid— The New and Improved Power Grid: A Survey," IEEE Communication Surveys and Tutorials, Vol. 14, No. 4, 2011, pp. 944-980.

25. M. A. Lisovich and S. B. Wicker, "Privacy Concerns in Upcoming Residential and Commercial Demand-Response Systems," Proceedings of the Clemson University Power Systems Conference, 2008.

26. D. M. Han and J. H. Lim, "Design and Implementation of Smart Home Energy Management Systems Based on Zigbee," IEEE Transactions on Consumer Electronics, Vol. 56, No. 3, 2010, pp. 1417-1425.

27. E. Negeri and N. Baken, "Smart Integration of Electric Vehicles in an Energy Community," Proceedings of the 1st International Conference on Smart Grids and Green IT Systems, Porto, 19-20 April 2012, pp. 25-32.

28. The Gaia Tool. http://www.phasetophase.nl/en_products/vision_lv_network_design.html

29. MON, "Mobility Research of the Netherlands," 2009. http://rijkswaterstaat.nl/

30. L. Wang, V. Devabhaktuni and N. Gudi, "Smart Meters for Power Grid—Challenges, Issues, Advantages and Status," IEEE/PES Power Systems Conference and Exposition (PSCE), Phoenix, 20-23 March 2011.

5

CHAPTER

EGYPTIAN WIDE AREA MONITORING SYSTEM (EWAMS) BASED ON SMART GRID SYSTEM SOLUTION

M.M. Eissa[1], Mahmoud M. Elmesalawy[2], Ahmed Soliman[2], Ahmed A. Shetaya[1] and Mahmoud Shaban[1]

[1] Department of Electrical Engineering, Faculty of Engineering - Helwan University, Cairo, Egypt

[2] Department of Electronics, Communications, and Computer Engineering, Faculty of Engineering - Helwan University, Cairo, Egypt

1. INTRODUCTION

Wide-area measurement systems (WAMS) in smart grid can be defined as a system that captures measurements in the power grid over a wide area and across traditional control boundaries, and then uses those measurements to improve grid stability and events through wide-area situational awareness and advanced analysis.Authors have achieved Wide Area Monitoring System (WAMS) based on Frequency Disturbance Recorders (FDRs) as a family of the PMUs deployed on Egyptian Power grid for mapping and visualization of all system parameters. The FDRs are deployed on live 220kV/500kV Egyptian grid system in cooperation with the Egyptian Electricity Transmission Company (EETC). The project is funded from the National Telecommunication Regulatory Authority (NTRA). The Egyptian

Wide Area Monitoring System (EWAMS) achieved at the Helwan University can gather information from many FDR units geographically dispersed throughout the boundary of the Egyptian power grid and data manipulated at a data center contains many servers at Helwan University. The Synchrophasor system with wide deployment of using (FDRs) phasor measurement units and high-speed communications to deliver and collect synchronized high-speed grid operating data, along with analytics and other advanced on-line applications will improve real-time situational awareness and decision support tools to enhance system stability. The EWMS is a good environment for many applications that can help the EETC to enhance the Egyptian Grid. This manual Guide explains the remote access for the EWAMS established at Faculty of Engineering through the EETC. The access will be for displaying the FDRs' parameters with different configuration through web service.

The growing global population is driving an even greater increase in the demand for electricity. Added to this, governments around the world are focusing on reducing carbon dioxide (CO2) emissions by increasing the utilization of renewable energy sources in the power chain. Today, existing grids are under pressure to deliver the growing demand for power, as well as provide a stable and sustainable supply of electricity. These complex challenges are driving the evolution of smart grid technologies. The greatest future challenge is the integration with the renewable energy resources and controlling of it to reflect the best impact on the massive energy production [1].

Optimal use of ageing assets strategy is as much determined by economic factors as it is by technical issues. Factors such as obsolescence need to be considered when determining future needs. Increasingly companies are turning to risk based assessments to provide a holistic approach. Any Asset Strategy should be implemented and supported by a range of policies covering process safety, maintenance and inspection, renewal, and competence. The smart power transmission networks are conceptually built on the existing electric transmission infrastructure. However, the emergence of new technologies (e.g new materials, electronics, sensing, communication, computing, and signal processing) can help improve the power utilization, power quality, and system security and reliability, thus drive the development of a new framework architecture for transmission networks [2]. The traditional power grid is unidirectional in nature. Electricity is often generated at a few central power plants by electromechanical generators, primarily driven by the force of flowing water or heat engines fueled by chemical combustion or nuclear power. In order to take advantage of the economies of scale, the generating plants are usually quite large and located away from heavily populated areas. The generated electric power is stepped up to a higher voltage for transmission on the transmission grid. The transmission grid moves the power over long distances to substations. Upon arrival at a substation, the power will be stepped down from the transmission level voltage to a distribution level voltage. As the power exits the substation, it enters the distribution grid. Finally, upon arrival at the service location, the power is stepped down again from the distribution voltage to the required service voltage(s) [3]. Fig. 1 shows an example of the traditional power grid [4].

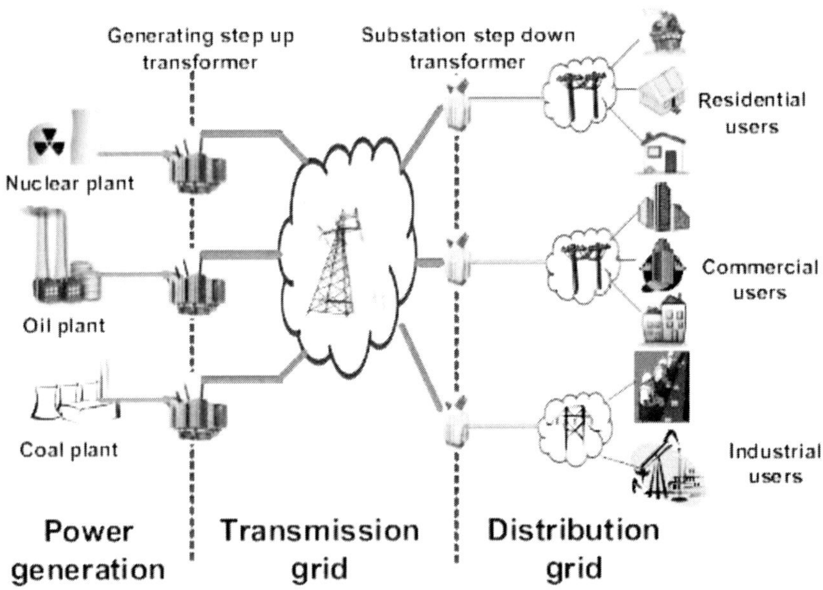

Figure 1. Traditional power grid [4].

The future grids can be regarded as an electric system that uses information, two-way, cyber-secure communication technologies, and computational intelligence in an integrated fashion across electricity generation, transmission, substations, distribution and consumption to achieve a system that is clean, safe, secure, reliable, resilient, efficient, and sustainable. This description covers the entire spectrum of the energy system from the generation to the end points of consumption of the electricity [5]-[6]. Fig. 2 shows a typical configuration for the future grid.

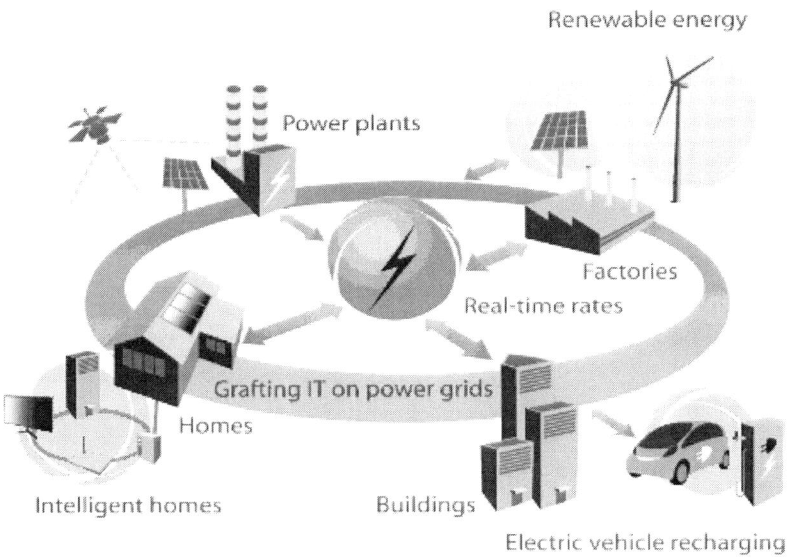

Figure 2. Future power grid [7].

The following Table 1 shows a comparison between the traditional and future power grid from different point of view.

Table 1. Comparison between Traditional and future power grid

Traditional power grid	Future power grid
Electromechanically	Digital
One way power flow	Multi-way power flow
One way communications	Multi-way communications
Few sensors	A lot of IED sensors
Low power quality	High power quality
Limited power flow control	Full power flow control (FACTS)
Manual monitoring	Self-wide area monitoring
Manual restoration	Self-Healing
Failures and blackouts	Adaptive and Islanding
Few consumer choice	Demand side management
No interconnected Renewable resources	Accept Renewable resources

2. EWAMS ARCHITECTURE

Egyptian Wide Area Monitoring System (EWAMS) is a 3G based wide area sensor network consisting of a special type of high precision family of the phasor measurement units (PMUs) and a central data management and processing system. EWAMS is a smart grid system designed to collect real-time synchronized frequency, voltage, and phase angle measurements at the transmission level 220kV/500kV of the power grid.

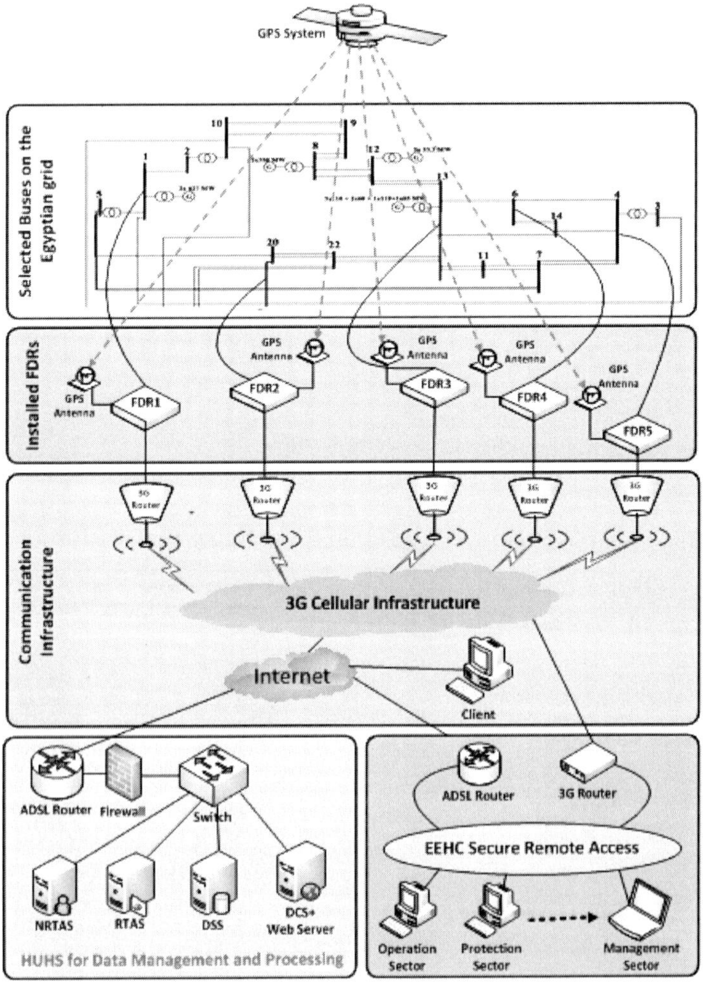

Sheme 1. EWAMS Architecture.

The structure of the EWAMS can be classified into five levels. The first level represents the placements of power stations on the power grid in which the sensors are installed. The second level contains the GPS enabled sensors that provide frequency, voltage magnitude, and voltage angle measurements. The third level is the communication infrastructure that provides the integrated wide area communication media for data measurements transmission. The fourth level is the remote data management and processing center that provides data gathering, storage, web service, post-disturbance analysis and other information management functions. The last level is the secure remote access connection for different EETC sectors and other remote clients. The following Figure shows the EWAMS architecture deployed for the Egyptian power grid. In order to discuss the unique characteristics of the EWAMS, the main building blocks of the EWAMS are discussed in more details in the following subsections.

The TCP and IP headers have numerous fields that are used to support the intended TCP and IP functionality as shown in Figure 6(a). In the proposed wide area measurement system, IP is used for delivering the different measurements collected by FDRs distributed over a wide area to the correct Data Concentrator Server (DCS) located at DMS while TCP is used to provide a reliable communication channels between FDR devices and the DCS.

The TCP connection between each FDR and the DCS is established through a three-way handshake process, ensuring that both FDR and DCS have an unambiguous understanding of the sequence number space. The operation of the connection is as follows:

a. The FDR sends the DCS an initial sequence number to the predefined destination port, using a SYN packet.

b. The DCS responds with an ACK of the initial sequence number and the initial sequence number of the FDR in a response SYN packet.

c. The FDR responds with an ACK of this DCS sequence number.

d. The connection is opened.

The operation of this algorithm is shown in following Figure. The performance implication of this protocol exchange is that it takes one and a half round-trip times (RTTs) for each FDR and DCS to synchronize state before any data can be sent. After the connection has been established, the TCP protocol manages the reliable exchange of data between FDRs and DCS. The existing UMTS mobile communication infrastructure is used to provide Internet access connection for FDRs. High speed packet access (HSPA) USB modems connected to 3G routers are used to provide the Internet access connection. Client server model is used to provide the communication between FDRs devices and DCS server. Each FDR device is act as a client and requesting to make a connection with the DCS server. The measurements data is transmitted from FDRs devices to the DCS server using a standard network protocols, Transmission Control Protocol (TCP)/Internet Protocol (IP).

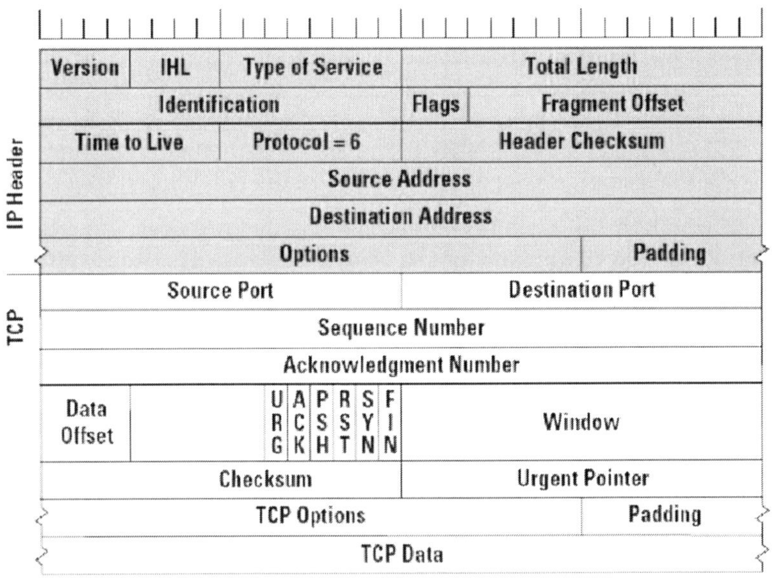

Sheme 2. The TCP/IP Datagram.

a. **EWAMS Sensors (FDRs)**

In EWAMS, the FDR is especially designed and implemented to have two power inputs. The first one is the single phase voltage (57.73VAC) which is taken from the 220/500 kV voltage transformer output in each power station and is used for estimating the three required parameters (frequency, voltage magnitude, and voltage angle). The other input is 220 VAC from power outlet which is used to supply the FDR device with the required power for its electronic circuit operation.

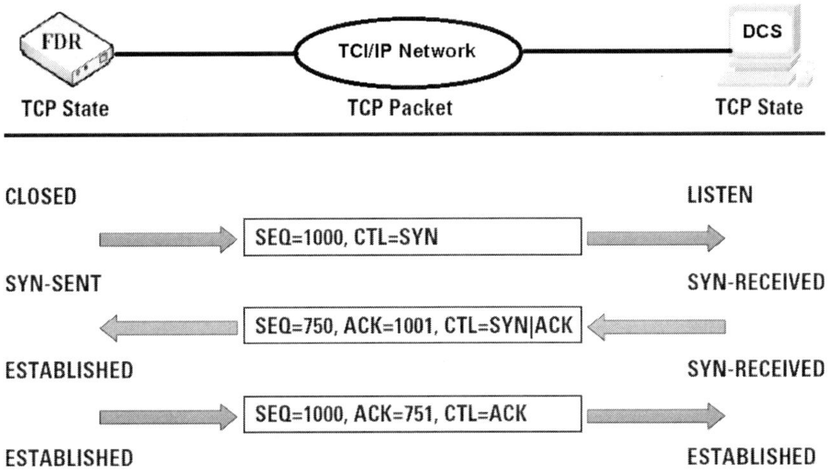

Sheme 3. TCP Connection establishment between FDR and DCS.

b. **Communication Infrastructure for EWAMS**

In EWAMS, FDR devices are distributed over a wide area, covering various locations within the boundary of the power system. The FDR devices are then connected to the remote data center through communication network. Public telecommunication infrastructure (Internet) will be used to provide the communication channels between FDRs and the data concentrator server (DCS) in the data center. The Data Concentrator Server (DCS) located at the Egyptian Wide Area Monitoring System "EWAMS" hosted at Helwan University. Transmission Control Protocol/Internet Protocol (TCP/IP) is used in building the system. The TCP connection between each FDR and the DSC is established through a three-way handshake process, ensuring that both FDR and DSC have an unambiguous understanding of the sequence number space. The FDR sends the DSC an initial sequence number to the predefined destination port, using a SYN packet. The FDRs transmits their measurements over the Internet to the DCS using the standard Internet Protocol suite TCP/IP. Since, the Internet Protocol doesn't guarantee any quality of service which it provides best effort delivery for data packets. The missed data at the DCS is expected as a result of a number of factors including packet drop due to network congestion, buffer overflow at the FDR or DCS, and corrupted packets rejected in-transit or faulty networking hardware.

3. HELWAN UNIVERSITY HOST SERVERS (HUHS)

The HUHS is data management and processing center operated by several dedicated servers. The logic behind decomposing the HUHS to several numbers of servers is to distribute the computation power which had the advantage of increasing the systems redundancy and reliability. The HUHS consists of four servers connected together through local area Network (LAN). HUHS can divided in terms of functionality to four servers, the data concentrator server (DCS), the real-time application server (RTAS), the Non-Real-time application server (NRTAS), the web server, and the data storage server (DSS). The main functions of each server are described below.

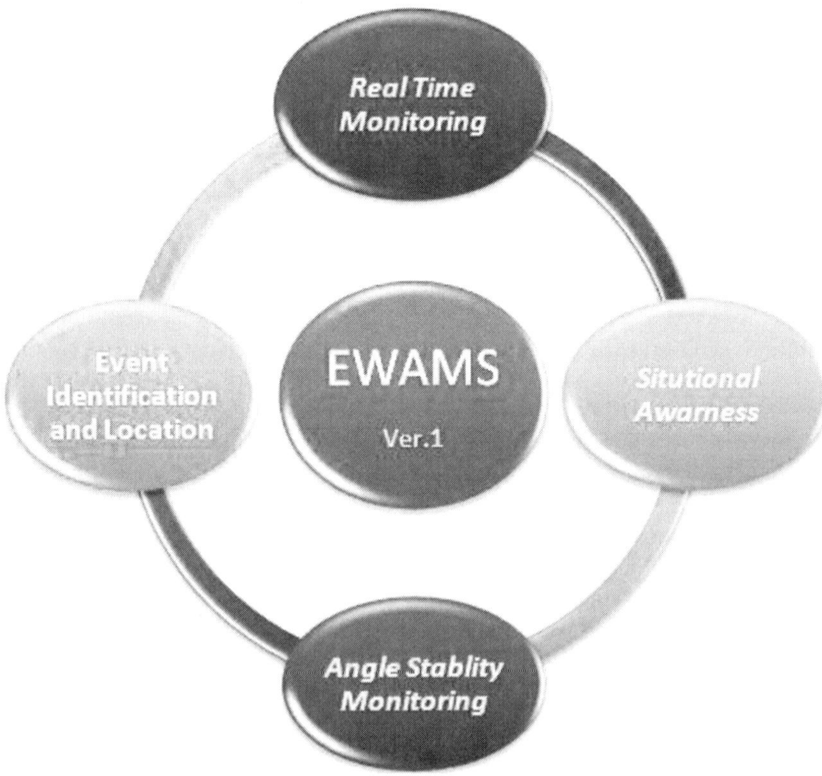

Figure 3. The four different applications in the EWAMS.

4. EWAMS CAPABILITIES

EWAMS includes many applications in real time mode. The system includes real time situational awareness in different mode for displaying different parameters. The EWAMS system includes also many vital applications. The different software EWAMS capabilities are summarized in four main applications, see Figure 3.

a. **Real Time Monitoring**

In this application, the system introduces the features of the real time monitoring. In this step, the EWAMS system displays many measurements as trends. All measurements are synchronized at same time using GPS timing system.

b. **Real Time Situational Awareness**

In this application, the system introduces the features of the real time situational awareness. The situational awareness application will display many measurements in color mapping way on the Single Line Diagram of the Egyptian power grid that help the operator in taking decisions in real time in case of situational awareness issued from one screen.

c. **Real Time Stability Monitoring**

The most effect way to predict the grid operating state is "the real time angle stability monitoring". In this application, the angles of the monitored buses reflect the state of the areas of the angles on the Egyptian grid to indicate any alerts during the angle divergence. This application will give the operator complete image about the state of the system to take fast actions.

d. **Event Identification**

The EWAMS system has the ability to detect events and define its type either generator tripping or load shedding. It also estimates the mismatch power amount that rejected from the grid and defines the area of the event. The measurements during this event could be visualized at any time and date for current event or any another historical events. This study partially completed and under construction now for remote access.

e. **Historical Data**

One feature of the EWAMS system is the possibility to access at any time and date for getting historical data. This data facility introduces transient analysis for any historical events on the grid, planning for future work on the grid through parameters monitoring, forecasting analysis, etc. This features as a remote access is under the construction but it is available on the EWAMS.

4.1. Main Features of the Real time monitoring and situational awareness

- The frequency mapping will define the islanded areas and how it effects on the rest of the grid.
- The angle mapping will define the oscillated areas and also shows the weakness area on the grid.
- The voltage mapping will define the most loaded area to help in load shedding selection criteria.

4.2. Procedures Steps for EWAMS Accessing

This part explains in more details an explanation for the operator about the possibility of EWAMS access.

In the welcome page the user will first go to the main menu through the following page

1. Open your internet browser using Internet Explorer, Chrome or Firefox...etc.
2. Enter the main webpage has following Internet Link in the browser
3. Write the Username and Password in space fields.
4. Press Login Button and then the program webpage is displayed

Figure 4 shows the main menu.

a. **Accessing the "MAIN MENU" of the EWAMS**

After logging in, the system will go to the monitoring menu. Figure 5 shows the EWAMS software system. As shown in the figure the menu contains:

Vertical Menu that contains:

- **Trends**
- **Maps**

- **SLD**
- **Applications**

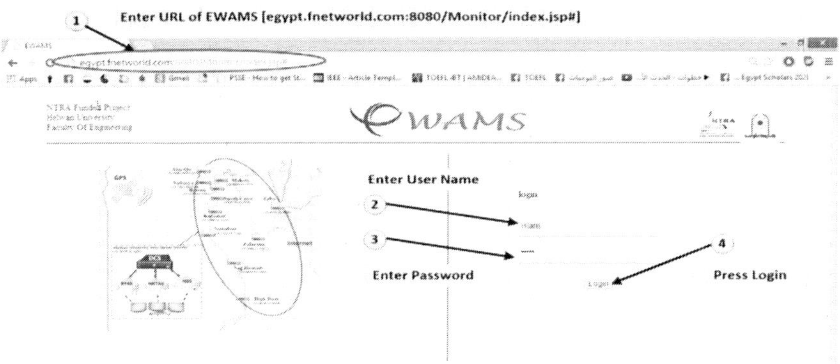

Figure 4. Main menu.

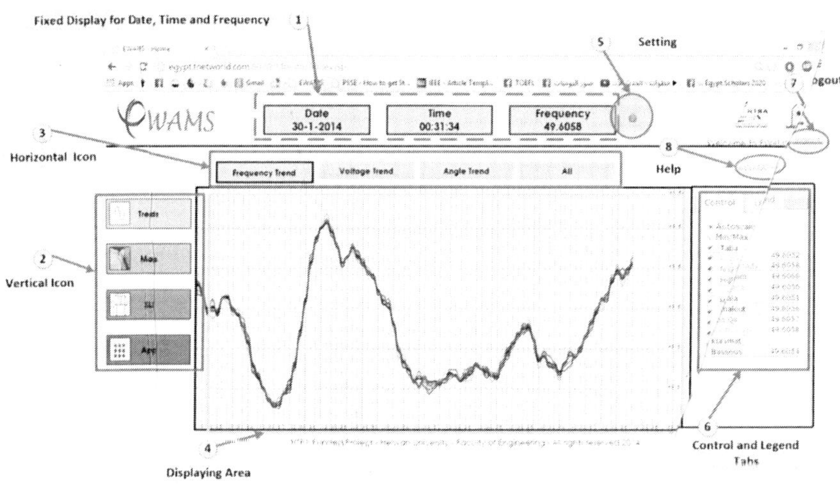

Figure 5. The nain menu of the EWAMS system.

Trends: are used for displaying the frequency, voltage, phase angles and all trends of the system in trends mode.

Maps: are used to display the mapping of the system parameters on the Egyptian map in MAP MODE

SLD: is used to display the system parameters on a single line diagram of the Egyptian grid

Applications: are used for displaying the "real time stability monitoring".

Figure 5 shows the horizontal menu. This menu appears based on the vertical menu.

Figure 5 shows also the three displaying menu for the Date, Time and Frequency of the HIGH dam that taken as the reference for the system.

In Figure 5, the right hand side of the main menu there are two tabs; "control" and "legend". The control tab displays a list of the FDR stations, and the corresponding frequency readings. There is the option to select the FDR stations and their readings. Also, there is a tab with "legend" that displays the color bar ranges of the frequency or angle. This tab will appear only in the "SLD" and "MAPs" on the vertical menu. Figure 5 shows also the "setting icon" that uses for selecting the reference frequency for displaying it on the vertical icons. It also used for determine the max and min frequency thresholds.

b. **Accessing "TRENDS"**

This part will deal with "real time monitoring". Figure 6 shows the EWAMS system when the operator wants to access the "Trends", he just clicks on the "Trends" icon". After clicking, a horizontal menu will be appeared that displays four icons "Frequency Trends", "Voltage Trends", "Angle Trends", and "All". Every icon will display the trends of the parameters located. Figure 6 shows also the frequency trends that will come from "Frequency Trends" icon in the horizontal mode.

Figure 7 shows the voltage trends, which will come after clicking the "Voltage Trend" in the horizontal menu. Figure 8 shows the angle trends that come from the clicking on the "Voltage Trends" in the horizontal menu. Figure 9 shows the all trends that come by pressing on "all" tab.

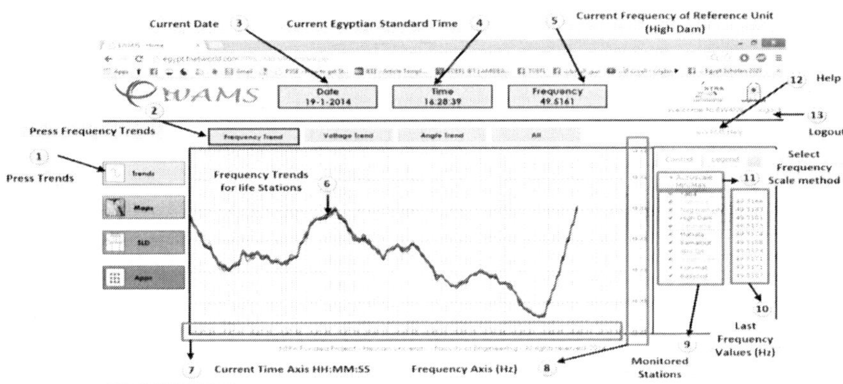

Figure 6. Accessing trends through EWAMS.

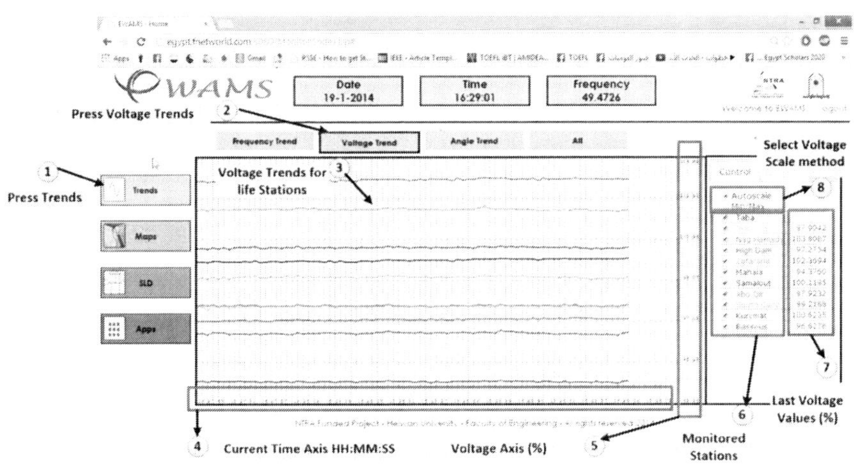

Figure 7. The voltage trends.

Figure 8 shows the right hand list of the control icon. In this list all FDR with their values will be appeared. There is the option to select the displayed FDR stations.

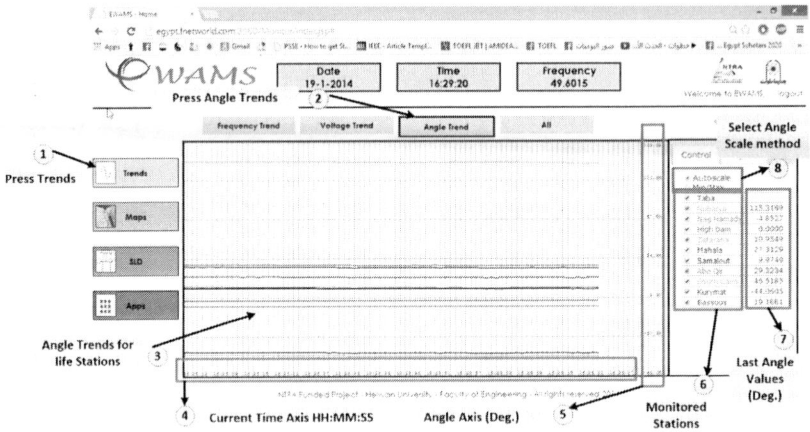

Figure 8. The voltage trends.

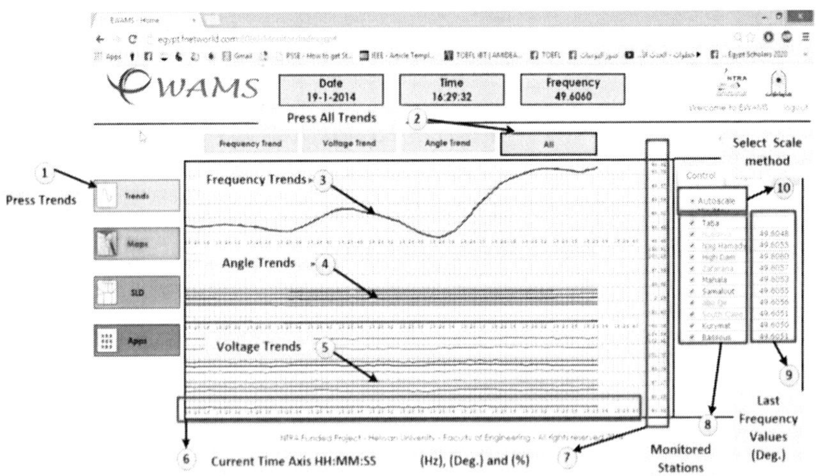

Figure 9. All trends.

c. **Accessing "MAPS"**

This part will deal with "Real Time Awareness". In this tab the frequency, angle and voltage will be displayed on a Map. To access this application the user must follow these steps:

To display Situational awareness by frequency, press "Maps" Tab, then press "Frequency Map" Tab given on the horizontal bar. Figure 13 shows the "Maps" and "Frequency map" for the frequency displaying. And so on for the "Voltage Map" and "Angle Map". On the right hand side a color code for the frequency readings that corresponding FDRs.

Figure 10. Control legend.

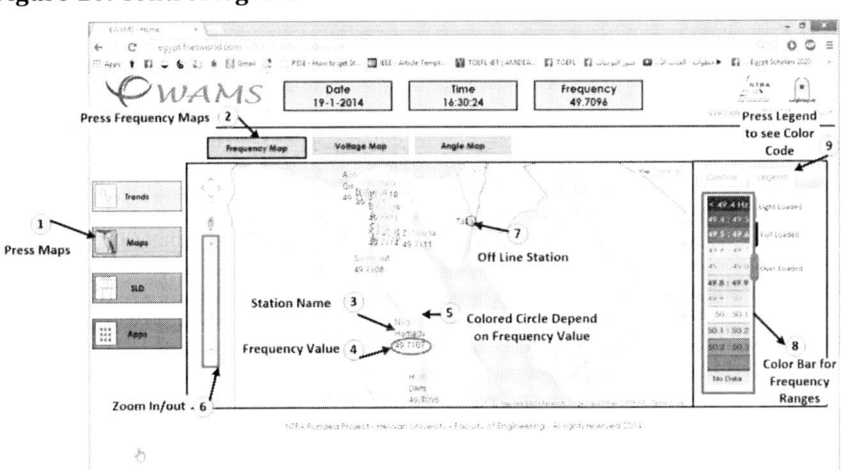

Figure 11. MAPS menu for the frequency displaying.

The FDR data can be used to enhance grid reliability for both real-time operations and offline planning applications [11]-[14], as listed below; the benefits of achieving such system will assist in:

- Wide-area situational awareness
- Real-time operations applications
- Frequency stability monitoring and trending
- Power oscillation monitoring
- Voltage monitoring and trending
- Alarming and setting system operating limits, event detection and avoidance
- Resource integration
- State estimation
- Dynamic line ratings and congestion management
- Outage restoration
- Operations planning

5. CONCLUSION

Wide Area Monitoring System on real 220kV/500kV Egyptian grid is achieved in cooperation with the Egyptian Electricity Holding Company in installing the devices. The EWAMS concerns the pioneers in the field of the smart grid wide area monitoring system. This system is a part of Smart grid system through transmission levels. On-line monitoring for different parameters is achieved. Many of new applications such as EWAMS Mapping and Visualization, Real time event detection and identification, Angle stability monitoring are applied based on remote access at the EETC. The system can satisfy many of the features for the Egyptian grid as Improved power system operation, Better use of existing equipment, Increased power transmission capacity, Lowered risks of power system instabilities, Improved power system planning, Less outage power and time, capability of integration renewable recourses, analysis and investigating many post-events.

6. ACKNOWLEDGEMENTS

The authors gratefully acknowledge funding and support from **National Telecom Regulatory Authority (NTRA)**, Egypt (http://www.ntra.gov.eg) to implement the network architecture proposed in this work. Also, I would like to thank Egyptian Electricity Transmission Company (EETC) for help in implementing the FDRs devices on real system. More details about the project progress can be followed on (www.helwan-ntra.com).

Special thanks for Prof. Yilu Liu (UT/ORNL Governor's Chair Professor_ University of Tennessee), USA and Prof. Hossam Gaber (University of Ontario Institute of Technology, UOIT), Canada

REFERENCES

1. PSGuarde, Improved Power System Performance through Wide Area Monitoring, Protection and Control, ABB, 2004.
2. A. Bose. Smart transmission grid applications and their supporting infrastructure. IEEE Transcations on Smart Grid, 1(1):11–19, 2010.
3. H. Farhangi., The path of the smart grid. IEEE Power and Energy Magazine, 8(1):18–28, 2010.
4. Fang, Xi; Misra, Satyajayant; Xue, Guoliang; Yang, Dejun, "Smart Grid — The New and Improved Power Grid: A Survey," Communications Surveys & Tutorials, IEEE, vol.14, no.4, pp.944,980, Fourth Quarter 2012.
5. Office of electricity delivery and energy reliability "Smart grid", Internet: http://www.oe.energy.gov/ smartgrid.htm, Accessed: August 2009.
6. Saint B, " Rural distribution system planning using Smart Grid Technologies", IEEE Rural Electric Power Conference, REPC '09, Page(s):B3 - B3-8, April 2009.
7. Miroslav M. Begovic," Smart Grid R&D Perspectives from the US" 3rd IEEE PES ISGT Europe, Berlin, Germany, October 14 -17, 2012.
8. M.D. Hadley; J.B. McBride; T.W. Edgar; L.R. O'Neil; J.D. Johnson, "Securing Wide Area Measurement Systems", U.S. Department of Energy, Office of Electricity Delivery and Energy Reliability, june 2006.

9. Jian Zuo, "The Frequency Monitor Network (FNET) Design and Situation Awareness Algorithm Development", Doctor of Philosophy, Blacksburg, Virginia U.S.A, March 2008.

10. Lang Chen; Markham, P.N.; Yilu Liu, "Wide-area dynamic model validation using FNET measurements," Innovative Smart Grid Technologies (ISGT), 2012 IEEE PES, vol., no., pp.1,7, 16-20 Jan. 2012.

11. M.M. Eissa, Yilu Liu, Mahmoud M. Elmesalawy, and Hossam Gabbar, "Wide Area Synchronized Frequency Measurement System with Secure Communication infrastructure for 500kV/220kV Egyptian Grid", IEEE International Conference on Smart Grid Engineering (SGE'12)-27-29 August, 2012-UOIT, Oshawa, Canada.

12. M.M. Eissa, Mahmoud M. Elmesalawy, Ahmed A. Shetaya, Ahmed H. Soliman, "Monitoring and Novel Applications of 220kV/500kV Egyptian Grid Parameters Using family of PMU based WAMS " 3rd international workshop for sustainable energy for all "transforming commitments to action", 22-24 February 2014, Christ university, Kengeri, India.

13. Elmesalawy, M.M.; Eissa, M.M., "New Forensic ENF Reference Database for Media Recording Authentication Based on Harmony Search Technique Using GIS and Wide Area Frequency Measurements," Information Forensics and Security, IEEE Transactions on, vol.9, no.4, pp.633,644, April 2014.

14. Moxley, R., Zweigle, G., Al-Mahrouq, A.: 'Wide-area measurements to improve generator and system modeling'. Innovative Smart Grid Technologies—Middle East (ISGT Middle East), 2011 IEEE PES Conf., 17–20 December 2011, pp. 1–7.

15. M.M. Eissa, Wael M. Fayek, Marwa M.A. Hadhoud, Mahmoud M. Elmesalawy, Ahmed A. Shetaya "Frequency/voltage wide-area measurements over transmission control protocol/internet protocol communication network for generator trip identification concerning missed data", IET Generation, Transmission & Distribution, 2014, vol. 8, no.2, pp. 290 – 300.

16. http://www.3gpp.org/technologies/keywords acronyms/99-hspa.

17. M.M. Eissa, Yilu Liu, Mahmoud M. Elmesalawy, and HossamGabbar, "Wide Area Synchronized Frequency Measurement System with Secure Communication infrastructure for 500kV/220kV Egyptian Grid", IEEE International Conference on Smart Grid Engineering (SGE'12)-27-29 August, 2012-UOIT, Oshawa, Canada.

18. M.M. Eissa, Mahmoud M. Elmesalawy, Ahmed A. Shetaya, "Smart Grid Frequency System on 220kV/500kV Egyptian Grid – Architecture and application" IAC 2014 International Conference on Industry Academia Collaboration, 3-5 March Fairmont Heliopolis Cairo-Egypt.

6

CHAPTER

INCREASING ENERGY EFFICIENCY BY REDUCING LOSSES AND PROMOTING VALUE PROPOSITIONS

Rune Gustavsson[1] and Leif Marcusson[2]

[1] KTH, Stockholm, Sweden

[2] Linneaus University, Kalmar, Sweden

1. INTRODUCTION

Energy is the base of our society and is the precondition of almost all human activities since at least a couple of centuries. Besides the benefits of energy usage we also have severe drawbacks related to negative effects on our environment. Manifested by, e.g., the climate change due to increased carbon dioxide emissions and scars and waste from coal mining and oil digging.

Energy efficiency has at least the following interpretations.

- Diminishing losses in generation, transmission and distribution of energy. It is estimated that we today have about 10% energy losses in the transmission and distribution grids of Alternating Current (AC) energy.
- Increasing performance of equipment and processes
- Increasing comfort and well being of humans

By increasing the performance of equipment we get 'more mileage' from each unit of energy. Better support tools for human activities in different areas, e.g., by smart robots, will increase the demand of energy as well as provide increased quality of work. Novel tools and techniques increase comfort and well being of humans in areas such as ´smart cities' and/or individual health-care and monitoring systems.

The following *conceptual energy triangle*, from AC technologies, supports discussions on different energy efficiency issues (Figure 1).

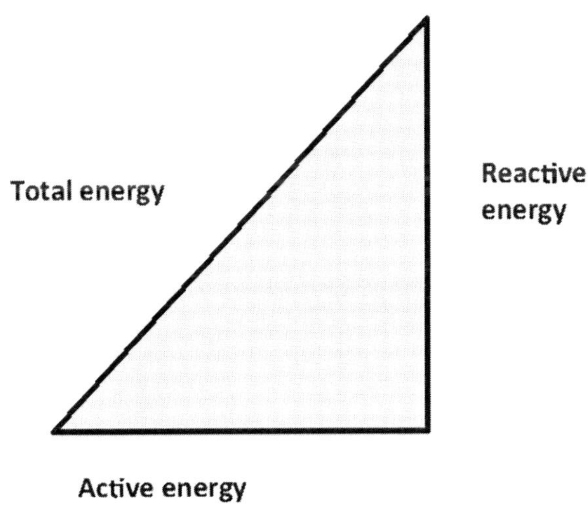

Figure 1. Conceptual energy triangle

In the case of AC generated power, *Active power* represents the power consumed by loads at customer's premises. That is, the power measured and paid for by customers. *Reactive powe*r is needed to continuously keep the electric balance of the power systems. Keeping this balance needs supporting equipment and real time monitoring and control provided by SCADA systems.

The following Figure 2, adopted from[1] - , captures the basic notations related to AC systems.

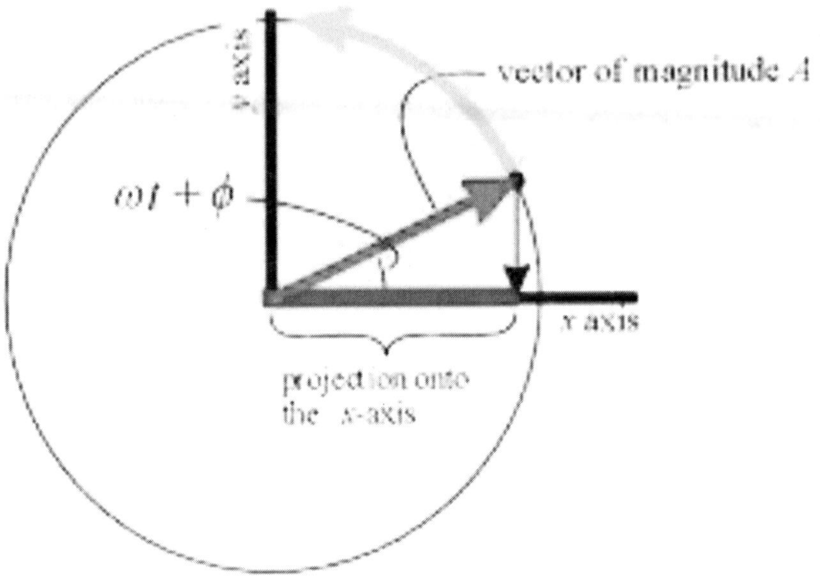

Figure 2. AC generation parameters at generators

Generators driven by an external energy source such as waterfalls typically generate AC power. The rotation in the generator causes AC generated by *interactions between magnetic and electric fields.* In Figure 2 this phenomenon is illustrated by rotations of the red total vector. Its projection on the x-axis is the *active power generated (KWs),* whence the projection on the y-axis is the *Reactive Energy (VARs).* The *sinusoidal waveform* of both projections have a phase angle of (ωt + Φ), where ω is the angle speed (radians/second), t denotes the time and Φ determines the Phase Factor (PF).

A sinusoidal alternating voltage applied to a *purely resistive* load results in an alternating current that is *fully in phase* (F = 0) with the voltage. However, in many applications it is common to also have a *reactive component* in the system, that is, due to the fact that the system possesses *capacitance, inductance,* or both. These electrical properties cause the current to *change phase* (Φ) with respect to the voltage: capacitance tending the current to lead the voltage in phase, and inductance to lag it.

Reactive power (measured in VARs) is present in a system containing reactive (induc-tive or capacitive) components and can be either produced or consumed by different load/generation elements. Though 'imaginary', the reactive power has great physical significance and is essential to the operation of the electrical system as a whole.

While the real power P is used to supply the energy required to perform actual work (such as running a motor), the *reactive power regulates the voltage* in the system. If the reactive power is too low, inductive loads such as transformers will be unable to maintain voltages necessary for the generation of electromagnetic fields, leading to a 'voltage collapse' that create blackouts.

Transmission line impedances also make it necessary to provide reactive power to maintain voltage levels necessary for active power to flow through. Therefore *reactive power is essential to move active power through transmission and distribution systems* to the customer. However if reactive power in a system is too high, there is *increased heat loss in transmission lines and loads as the current flowing through the system is much higher*, creating a potentially hazardous breakdown situation. The power factor (PF) of a load tells us what *fraction* of the *apparent power* (red in Figure 2) is in the form of real power and performs actual work. A high power factor is desirable since it minimizes the amount of reactive power needed by the load, reducing heat losses and maximizing efficiency (Section 2.3 and Section 5).

In the case of vehicles the *total energy* corresponds to the distance travelled whence reactive energy corresponds to energy needed to overcome different kinds of resistance (wind, road, engine) during the journey. The demand from customers to have more energy efficient vehicles has forced manufacturers to improve the efficiency by reducing the need of total energy in driving vehicles.

The customer costs are thus, as we know, different in those two cases. In the electrical power case the customer only pays for active energy, whence reactive energy has implicit costs for the end customer and is, at present, provided by grid operators. The two energy cost models; total costs versus partial cost paid by customers, have led to different focus of transport and energy system developments [Section 6].

The total cost model has led to customer driven demands of increased energy efficiency in design and maintenance of vehicles and a transport market with transparent total cost of ownership as well as clear roles of public and private stakeholders.

In AC energy systems the increased demands by authorities to include vast amounts of *Renewable Energy Sources* (RES), e.g., by wind farms or solar cells have stressed the transmission and distribution system abilities to keep the energy balance. That is, increased the costs for maintaining necessary (not billable) reactive power. The *only incentive* for customers of emerging Smart grids, however, so far is to save *active* energy.

In short, in order for future Smart grids to really take off to meet the societal expecta-tions we have to *reconsider current cost and revenue models (CMs and RMs)* to meet the emerging market demands, not the least from future customers [Section 5].

Energy as such has different interchangeable forms and sources. We focus on electricity markets, as it is arguably the most versatile form of energy.

The remainder of the chapter is organized as follows. In Section 2 *Introduction to Smart grids* we outline the motivations and approaches towards Smart grids. We specifically address challenges such as:

- Estimated increased demands of energy
- Meeting environmental challenges
- Reducing losses in power generation and distribution
- Architectural principles of Smart Grids

The *Smart Grid Architecture Model (SGAM)* supports design and implementation of different energy models. The business view of those can be analysed and modelled using the complementary *Business Model Framework eTransit* described in Section 4 *eTransit Framework.*

Key components of the eTransit framework are *Value Propositions* (VPs) and *Revenue Models and Cost Structures* (RMs). A given *Business Case* can be mapped onto the Layers of the SGAM Framework identifying *relevant stakeholders and infrastructures* to deliver the chosen services related to the selected VP. Issues related to *coordination and monitoring* of tasks and stakeholders are handled by design and implementation of *Service Level Agreements* (SLAs). Identification, design, implementation and monitoring of SLAs are addressed in Section 3 *Trustworthy Service Level Agreements.* Specific challenges are related to *interoperability,* assuring that stakeholders have a *shared understanding* of the monitored state of operations, and that *high-level business concepts are supported in real time while maintaining the energy balance* of the system. To specifically support *empowerment of customers* we are addressing issues related to *Cognitive Interaction Patterns* in Section 4.1.

We have now a necessary toolbox to address the issues raised earlier. That is, suitable Cost and Revenue models to meet future market demands of

Smart grids. This challenge is the topic of Section 5 *Suitable Cost and Revenue models for future Smart grids.* We exemplify our findings by introducing:

- Power Quality markets (Section 5.1)
- Cutting losses in power distribution grids (Section 5.2)
- Smart Home services enabled by Service Level Agreements (Section 5.3)

The Chapter concludes with Section 6 *Conclusions and Future Work* and Section 7 *References.*

2. INTRODUCTION TO SMART GRIDS

Current energy systems are under preasure of three challenges. *Firstly*, they must meet the expected increasing demands on energy globally (Section 2.1). *Secondly*, they must meet the demands of decreasing the negative effects on our environment (Section 2.2). Here we must, however, establish better models for evaluating and comparing *different approaches. Thirdly*, they have to be more efficient (Section 2.3). To meet those challenges the concept of *Smart grids* have been introduced (Section 2.4)

2.1. Estimated increased demands on energy

There are several prognoses of future energy demands. A recent one is provided by *U.S. Energy Information Administration* [1] In their *International Energy Outlook (IEO) 2013* the following Figure 12 is presented. According to the IEO2013 Reference case, world energy consumption has increased from 524 quadrillion (10^{15}) Btu (Brittish thermal unit) in 2010 to expected 630 quadrillion Btu in 2020 and 820 quadrillion Btu in 2040, hence a 30-year estimated increase of 56 percent. More than 85 percent of the increase in global energy demand from 2010 to 2040 occurs among the developing nations outside the *Organization for Economic Cooperation and Development* (non-OECD), driven by strong economic growth and expanding populations. In contrast, OECD member countries are, for the most part, already more mature energy consumers, with slow anticipated economic growth, and little or no anticipated population growth.

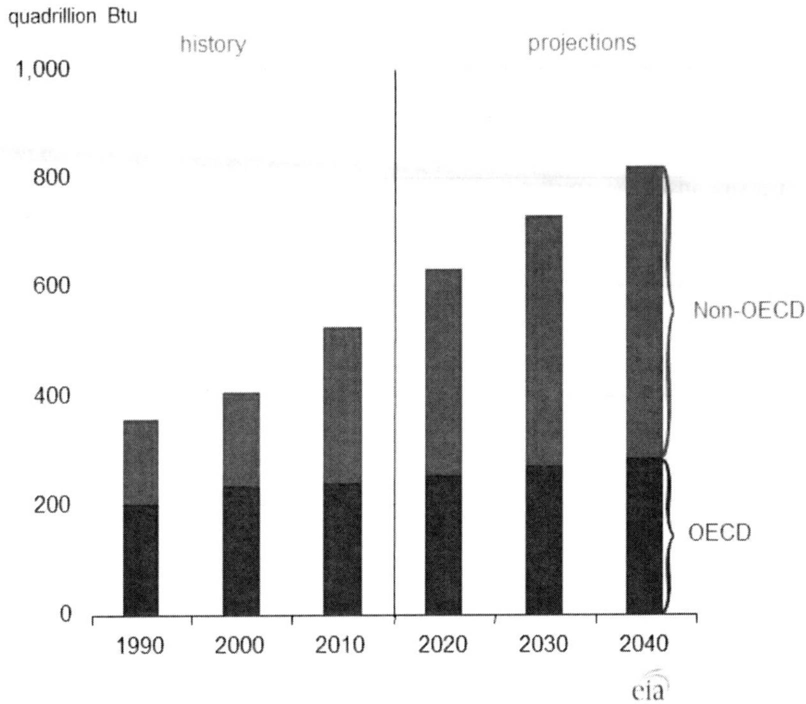

Many economic and geopolitical circumstances add considerable uncertainty to any long-term assessment of world energy markets. Currently, there is a wide variation in the economic performance of different countries and regions around the world. In the United States and Europe, short- and long-term debt issues remain largely unsolved and are key obstacles of future growth.

2.2. Meeting the environmental challenges

Governments in both Europe and the United States have implemented CO_2 emission policies of the following three types:

1. *Renewable Incentives*. Renewable initiatives are generally not available for non-carbon initiatives such as nuclear, large-scale hydro or gas combined cycles. Yet results from a recent report, below, demonstrate clearly that these three alternatives are far more cost effective per MW or capacity in reducing carbon dioxide emissions than wind or solar. In both the United States and Europe, there is, however, political

opposition to all three of these alternatives on environmental and safety grounds, despite their cost-effective superiority in reducing carbon dioxide emissions

2. *Carbon Trading Systems.* There is two generally recognized methods of introducing a price for carbon dioxide emissions: (1) a *carbon tax*, or (2) a *cap-and-trade system*[2] - for enforcing lower carbon dioxide emissions. The price of carbon emissions on the *European Trading System* (ETS) reached a peak of 30 euros in 2006 but was settled at 5 euros 2013. The reduction in the ETS carbon emission price, along with increases in the price of natural gas in Europe has *made coal more attractive* as an energy source

3. *Tighter Regulations.* The United States has failed to adopt a national carbon emission trading system because of political opposition in the U.S. Congress. However, the U.S. *Environmental Protection Agency* (EPA) is planning tighter regulations under the Clean Air Act of 1990. The proposed regulations would, in effect, have same of the same effects on coal-fired plants as a carbon price

It is likely to be *far less costly* to achieve reductions in carbon dioxide emissions through a effective carbon trading system that allows the market to determine the most effective way to reduce emissions rather than through establishment of EPA standards for emissions.

To address the environmental challenges we must *model and analyse* different aspects of energy production. The most common method for comparing cost of different electricity technologies is to compute the *Levelized Energy Cost* (LEC[3] -). It can be defined in a single formula as:

$$LEC = \frac{\sum_{t=1}^{n} \frac{I_t + M_t + F_t}{(1+r)^t}}{\sum_{t=1}^{n} \frac{E_t}{(1+r)^t}}$$

where;

- LEC = Average lifetime levelized electricity generation cost
- I_t = Investment expenditures in year t
- M_t = Operations and maintenance expenditures in the year t
- F_t = Fuel expenditures in year t
- E_t = Electricity generation in the year t

- r = Discount rate
- n = Life of the system

However, there are convincing arguments that levelized costs are not appropriate for *ranking technologies*. The trouble is that levelized costs do not take account of the costs of intermittency. Wind power is not generated on a calm day, nor solar power at night, so conventional power plants must be kept on standby—but the *associated costs* are not included in the levelized cost of renewables. Electricity demand also varies during the day in ways that the supply from wind and solar generation may not match, so even if renewable forms of energy have the same levelised cost as conventional ones, the value of the power they produce may be lower. In short, levelized costs are *poor at comparing different forms* of power generation.

In a recent report from Booking; *The Net Benefits of Low and No-Carbon electricity Technologies,* different approaches supporting comparisons of different forma of power generation are introduced and discussed [2]. The report examines five different low and non-carbon electricity technologies and present the net benefits of each under a range of assumptions. It estimates the costs per megawatt per year for wind, solar, hydroelectric, nuclear, and gas combined cycle electricity plants. To calculate these estimates the paper uses a methodology based on *avoided* emissions and *avoided* costs. Rather than comparing the more prevalent levelized costs.

The following three findings are reported:

1. *First* – assuming reductions in carbon emissions are valued at €50 per metric ton and the prize of natural gas €16 per million Bty or less – Finding: nuclear, hydro, and natural gas combined cycle have far more net benefits than either wind or solar. This is the case because solar and wind facilities suffer from very high capacity cost per megawatt, very low capacity factors and low reliability, which result in low avoided emissions and low avoided energy cost per dollar invested.

2. *Second,* low and non-carbon energy projects are most effective in avoiding emissions if a price for carbon is levied on fossil fuel energy suppliers. In the absence of appropriate price for carbon, new non-carbon plants will tend to displace low-carbon as combined cycle plants rather than high-carbon coal plants and achieving only a fraction of the potential reduction in carbon emissions. The price of carbon should be high enough to make production from gas-fired plants preferable to production from coal-fired plants, both in the short term, based on

relative short-term energy costs, and in the longer term, based on relative energy and capacity costs combined.

3. *Third,* direct regulation of carbon dioxide emissions of new and existing coal-fired as proposed by the U.S. Environmental Protection Agency, can have some of the same effects as a carbon price in reducing coal plant emissions, both in the short term and in the longer term as old, inefficient coal plants are retired. However, a price levied on carbon dioxide emissions is likely to be a less costly way to achieve reduction in carbon dioxide emissions.

The findings of the report is summarized in the following Figure 3 from *The Economist* July 26th – August 1st 2014, page 20.

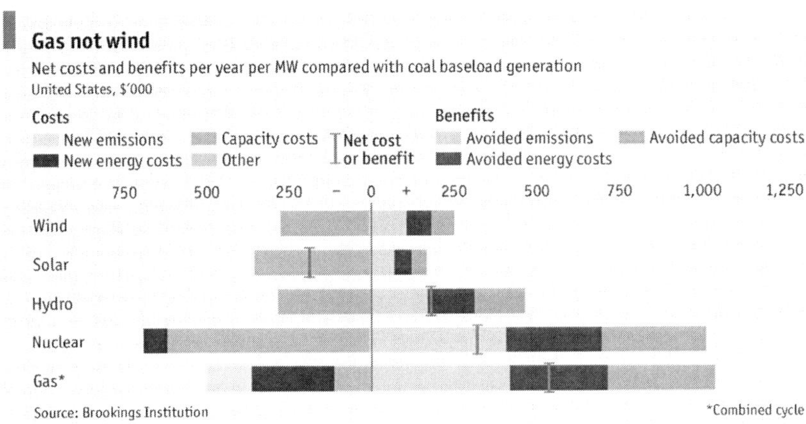

Figure 3. Cost-benefit analysis of different energy sources

If all the costs and benefits are added up using the model given in the report, solar power is by far the most expensive way of reducing carbon emissions. It costs $189,000 to replace 1MW per year of power from coal. Wind is the next most expensive. Hydropower provides a modest net benefit. But the most cost-effective zero-emission technology is nuclear power. The pattern is similar if 1MW of gas-fired capacity is displaced instead of coal. And all this assumes a carbon price of $50 a tonne. Using actual carbon prices (below $10 in Europe) makes solar and wind look even worse. The carbon price would have to rise to $185 a tonne before solar power shows a net benefit.

There are, of course, all sorts of reasons to choose one form of energy over another, including emissions of pollutants other than CO_2 and fear of nuclear accidents. Still, the findings have profound policy implications. At the moment, most rich countries and China subsidize solar and wind power to help stem climate change. Yet this is the most expensive way of reducing greenhouse-gas emissions. Meanwhile Germany and Japan, among others, are mothballing nuclear plants, which (in terms of carbon abatement) are cheaper. The implication of the research reported is clear: *governments should target emissions reductions from any source rather than focus on boosting certain kinds of renewable energy!*

2.3. Reducing losses in power generation and distribution

Power generated in power stations pass through large and complex networks of transformers, overhead lines, cables and other equipment before reaching the end users. It is fact that the Unit of electric energy generated by Power Station does not match with the units distributed to the consumers! Some percentage of the units is lost in the transmission and distribution networks. This difference in the generated and distributed units is known as *Transmission and Distribution (T&D) losses*[4] - .

- Transmission and Distribution losses, of AC systems, are not paid by users. They only pay for the received power (part of the Active energy, Figure 1).
- The Distribution sector is considered as the weakest link in the entire power sector efficiency measurement chain. Transmission Losses is approximate 17% while Distribution Losses is approximate 50% of the generated AC power.

There are two identified types of T&D losses. The losses are identified as *Technical losses* or *Non-technical losses.*

2.3.1. Technical losses[5] -

The technical losses are due to energy dissipated in the conductors, equipment used for transmission Line, Transformer, sub- transmission Line and distribution Line and magnetic losses in transformers. Technical losses are normally 22.5%, and directly depend on the network characteristics and the mode of operation. The major amount of losses in a power system is in primary and secondary distribution lines. While transmission and sub-transmission lines account for only about 30% of the total losses. Therefore the primary and secondary distribution systems must be properly planned to ensure within limits. Losses are inherent to the distribution of electricity and cannot be eliminated [3].

Variable losses vary with the amount of electricity distributed and are, more precisely, proportional to the square of the current. Consequently, a 1% increase in current leads to an increase in losses of more than 1%. Between 2/3 and 3/4 of technical (or physical) losses on distribution networks are variable Losses. By increasing the cross sectional area of lines and cables for a given load, losses will fall. This leads to a direct trade-off between cost of losses and cost of capital expenditure. It has been suggested that optimal average utilization rate on a distribution network that considers the cost of losses in its design could be as low as 30 percent.

2.3.2. Main reasons for technical losses

The main reasons for technical losses are:

- Lengthy Distribution lines
- Inadequate Size of Conductors of Distribution lines
- Installation of Distribution transformers away from loads centers
- Low Power Factor (PF) of primary and secondary distribution systems
- Bad Workmanship
- Feeder Phase Current and Load Balancing

One of the easiest loss savings of the distribution system is *balancing current* along three-phase circuits. Feeder phase balancing also tends to balance voltage drop among phases giving three-phase customers less voltage unbalance. Amperage magnitude at the substation doesn't guarantee load is balanced throughout the feeder length. Feeder phase unbalance may vary during the day and with different seasons. Feeders are usually considered "balanced" when phase current magnitudes are within a specified interval. Similarly, balancing load among distribution feeders will also lower losses assuming similar conductor resistance. This may require installing additional switches between feeders to allow for appropriate load transfer.

Power Load Factor (PF) [Figure 2] has several effects on losses. Power consumption of Customer varies throughout the day and over seasons. Residential customers generally draw their highest power demand in the evening hours. Same commercial customer load generally peak in the early afternoon. Because current level (hence, load) is the primary driver in distribution power losses, keeping power consumption more level throughout the day will lower peak power loss and overall energy losses. Load variation is denoted load factor and it varies from 0 to 1. *Power and energy losses are reduced by raising the load facto*, which, evens out feeder demand variation throughout the feeder.

The load factor can be increased by offerings to customer *time-of-use rates*. Companies use similar pricing power to influence consumers to shift electric-intensive activities during off-peak times (such as, electric water and space heating, air conditioning, irrigating, and pool filter pumping).

With financial incentives, some electric customers are also allowing utilities to interrupt large electric loads remotely through radio frequency or power line carrier during periods of peak use. Utilities can try to design in higher load factors by running the same feeders through residential and commercial areas.

To decrease T&D losses we need to be able to optimize the grid structure dynamically. To that end we need to implement a *suitable information processing system by monitoring relevant system parameters and take appropriate actions*. This is one of the goals of future Smart grids! Obviously, in reducing losses we also improve the environmental effects of power production.

In Section 5 we give an example of reducing losses in power distribution by installing 'smart equipment' and by empowerment of customers.

2.4. Architectural principles of smart grids

Smart Grids are in focus of several international R&D past and present efforts since at least a decade. The EU *Smart Grids Technology Platform* [4] has recently published *SmartGrids SRA 2035* outlining R&D efforts to be met 2035 [5]. Smart Grids is a well-known metaphor for future power grids. However, the meaning, or semantics of the concept has, naturally, changed due to increased understanding of the inherent complexities of the subject matter.

The driving forces behind the efforts on Smart Grids include:

- Demands of increased energy efficiency to meet the 20-20-20 EU goals
- Demands of integrating new energy sources such as *Distributed Energy Sources* (DES) and *Renewable Energy Sources* (RES) in a massive way into generation and transmission grids of future energy systems.

- Establishment of a de-regulated customer oriented energy market, including new types of energy based service markets.

The transition from today's mostly hierarchical power grids towards tomorrow's Smart Grids poses several challenges to be properly addressed and harnessed.

Among the challenges we have systems of systems requirements such as:

- Supporting information protection and security, as well as maintaining trustworthiness of services and systems by involved stakeholders in different use cases.
- Supporting interoperability and flexibility of involved systems.

Introduction of 'Smartness' of systems, such as classical power system, requires some new views and designs of systems to be implemented in a proper way. Some decades ago those systems were mainly hierarchical stovepipe hardcoded systems with SCADA systems monitoring and controlling the generation, transmission and distribution of energy. The SCADA systems enabled operators to maintaining agreed upon Quality of Service (QoS) to the sockets at customer's premises.

A first step to allow for flexibility across and between stovepipe systems is to *decouple* hierarchical systems into open *Service- Oriented Systems* (SOS), and to introduce *open reconfigurable interfaces* between the components. That is, standard interfaces sup-ported by tools and methods to *inspect and assess data-flows* across those interfaces (Section 4.1). Those methods also allow design and implementation of *defence-in-depth* to carter for increased system- and data security [6, 7].

A second step is to *decouple and supplement* the traditional SCADA and Protection systems into SOS and to introduce *dedicated* Information Processing Systems supplementing the SCADA systems. Its purpose is to Monitor and Control information flows supporting integration of new types of energy sources and flows of business transactions. In effect, future Smart Grids will consist of three *interconnected* types of systems:

- Systems supporting *Energy management*
- Systems supporting *information management*
- Systems supporting *Business management*

The first two types of systems are specified in Figures 4 – 7. However, the related Busi-ness management systems have so far *not been explicitly specified* to the same extent. This complementary view on Smart grid has also to be in focus in order to assess specific challenges related to *market models* of The Smart grids. We introduce to that end the *eTransit Business Model Framework* in Section 4! This allows us to specifically model *user centric case models* where we can illustrate *interdependence between all three views* of Smart grids (Section 5). In short, introduction of the eTransit Framework complements the Smart Grid Architecture Models, below, in a *crucial* aspect!

Different architectures of Smart grids have been proposed. For instance the *Smart Grid Architecture Model (SGAM)* outlined in Figure 4 and Figure 5 below [8].

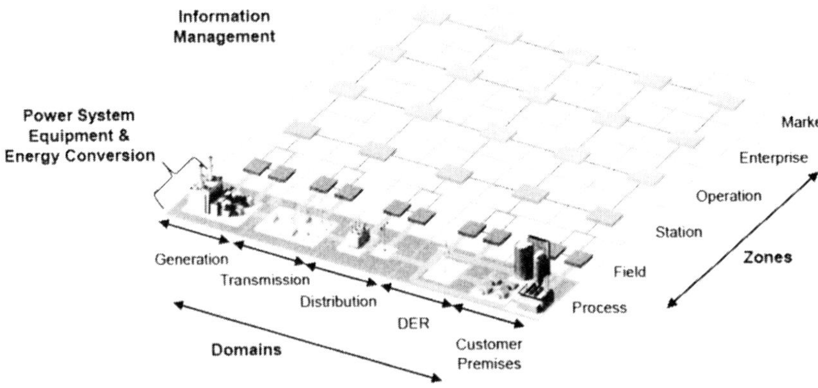

Figure 4. Smart Grid plane of SGAM

The Smart Grid plane consists of a grid with two axes of *Domains* and *Zones*. The Domain consist of components related to *Power System Equipment & Energy Conversion*, e.g., *Generation*, *Transmission* and *Distribution* as in classical power systems. However, and important *added* feature of Smart grids are the explicit components of *DER and Customer Premises*. The Zones consists of the complementary axis to Domains, that is, generalizations of the classical components *Processes, Field, Station, Operation* complemented with *Enterprise* and *Market*.

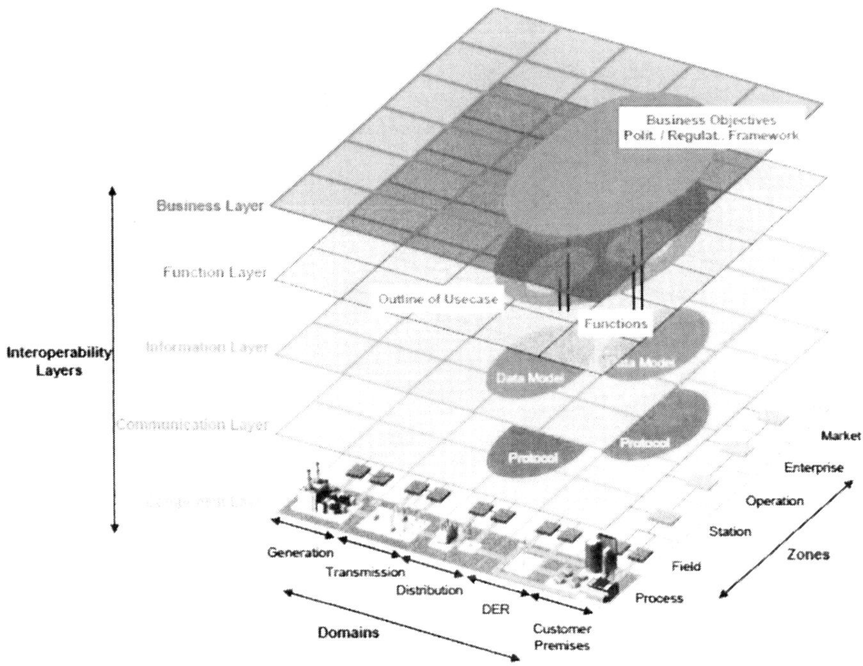

Figure 5. ayers of SGAM framework

The purpose of this Smart Grid plane is to implement and support the related *Information Management* of the individual and integrated three Infrastructures of Smart grids mentioned above. In order to carter for this very complex task we need to introduce a third axis, vertical and above, the Smart Grid plane. This axis of *Interoperability layers* is depicted in Figure 5, above.

The Interoperability layers are from bottom to top: *Component, Communication, Information, Function,* and *Business* layers. In Figure 5 we also have depicted a *projection from a Business Case* with stated objective down onto the SGAM framework cube.

In order to proper address issues related to Interoperability, the *Interoperability Framework* proposed by *GridWise Architecture Council,* (Figure 6), is incorporated in SGAM.

Interoperability issues are treated in some detail in Section 5.3 *Smart Home Services enabled by Service Level Agreements.*

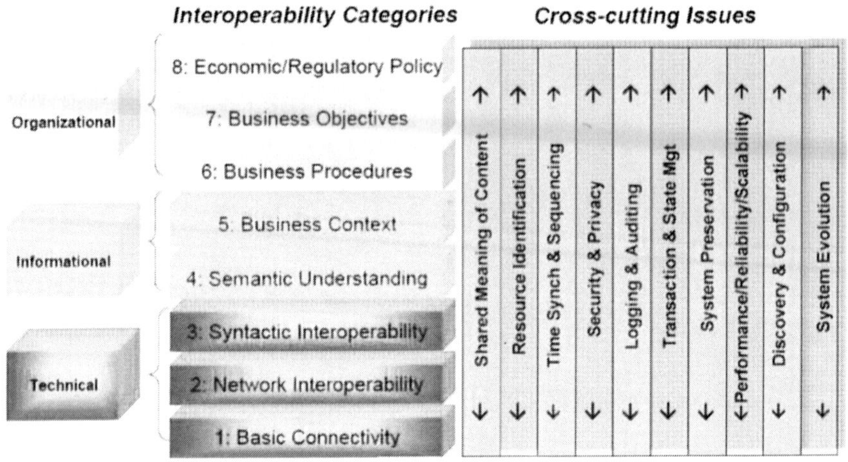

Figure 6. GWAC Interoperability Framework with a layered set of Categories and non-functional Cross-cutting Issues

In a classical power system the relevant Doman Component were *hierarchically integrated* with relevant zones into a *closed system* delivering and billing power to the customer according to relevant regulations and tariffs. The customers were just loads or ´two holes in the wall´. The necessary monitoring, control, and billing were supported by SCADA systems.

Transitions towards flexible and resilient Smart grids pose several basically hitherto unknown challenges. We know that power system engineering has been very successful during the last century in designing and maintaining the most complex human engineered infrastructures worldwide today. However, from Figure 5 and Figure 6 we can infer that this well renowned competence has to be *suitable complemented* with other competencies to meet the new challenges. Candidates include areas of computation and information sciences as well as competencies from economics and behavioral sciences and from psychological insights (Section 4.1).

Furthermore, to meet the societal demands we have to develop *suitable roadmaps* guiding political decisions related to procurement and evaluation towards understanding of relevant *socio-technical-economic* issues [5].

An illustrative example is the current focus of exploring RES to lower the CO_2 emissions causing the expected global warming. As we know those efforts are heavily subsided by governments worldwide. However, *the economic and environmental end results of those efforts are still not sufficiently understood and need more modeling and analysis* (Section 2.2).

In this chapter we focus on *Empowerment of customers* as a mean to increase energy efficiency in Smart grids as well as promoting environment protection. We know that introduction of massive amounts of DER and RES, e.g., by wind farms, will demand complementing by reactive power to keep the needed energy balance of AC systems. At the moment R&D efforts on RES and DER are heavily subsidized in several countries. This might lead to unsustainable market in the near future. To that end, there are ongoing discussions to implement some full-cost services in *Ancillary Service Markets* on:

- Operation Reserve
- Frequency Control
- Reactive Power and Voltage Control
- Short – Circuit Capacity
- Voltage Quality
- Black-Start
- Island Operation of the Grid
- Inertia
- Stability

Those markets are necessary to address, for instance, reducing losses in power generation and distribution (Section 2.3) and will be further discussed in Section 5.

Empowerment of customers might also support new *energy based services* that will increase several kinds of Energy efficiency aspects mentioned above (Section 6).

3. TRUSTWORTHY SERVICE LEVEL AGREEMENTS

Present power systems have a straightforward business model. The generation and distribution is in accordance with rules and *regulations set up and monitored by regulators.* The customers sign standardized documents and are billed based on the KWh consumption measured by meters installed at customer site and owned by a distributor. In liberated markets the bills *separates the costs into two parts* for (active) energy and for the corresponding use of network (net tariffs). The customer can choose between different energy providers and/or price incentives aiming at matching the supply/demands at peak hours (Section 2.3). The customers can also be compensated according to fixed rules in case of, for instance, blackouts.

Obviously, this simple business model cannot meet the requirements of future Smart grids! The standardized contract has to be replaced with a more flexible and market oriented concept.

We will, to that end, identify and establish a *Trustworthy Service Level Agreement* (SLA) based on a *selected Use Case* (UC) at the Business Layer in Figure 5.

Projections of a selected use cases onto lower Layers of SGAM (Figure 5) will determine:

- Relevant *stakeholders* and their concerns related to the chosen use case
- Identification of *Key Performance Indicators* (KPIs) to be monitored
- Identification of relevant functions (*business services*) at the Function layer
- Identification of supporting *information models* and *information exchange mechanisms* at the Information Layer
- Selecting relevant *communication systems and protocols* at the Communication Layer
- *Connecting and interfacing* the communication layer with selected components from the Component Layer

This process will design a business case as a *set of information processing boxes* from the Function-, Information-, and Communication Layers connecting the Use Case with components from the Component Layer. The information processing boxes (*information services*) have to be implemented and loosely coupled, to allow inspection and re-use, into a reliable information system that meet *Interoperability* (semantic) requirements as well as trustworthiness requirements from stakeholders in the use case.

The purpose of a SLA is to *coordinate and monitor* the business services and information services to meet the agreed upon requirements of the use case.

We have proposed a validated methodology to negotiate and establish suitable SLAS for Smart grids [9].

From the analysis above it follows that a use case can be modeled as a *set of vertical and horizontal workflows* (*business channels*) built from *chains of services* and *connecting components* from the Component Layer *with stakeholders* at the Business Layer (Section 5). To support analysis and modeling of the relevant *flexible* information systems we need a set of *meta models* as those defined by the eTransit Framework in the next section, Section 4.

The following Figure 7 captures the contexts of SLAs. The design and implementation of SLAs are based on *Drivers and opportunities* formulated as *requirements* related to Use cases. Those requirements have to be *matched against affordances* from the infrastructures.

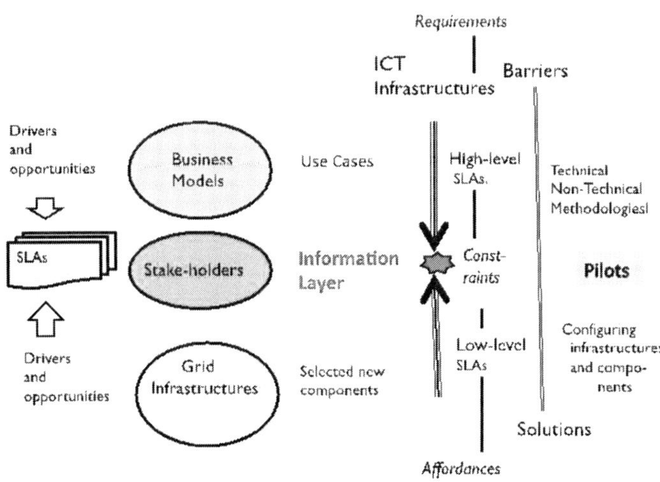

Figure 7. SLA context as coordination of bundles of high-level and low-level SLAs

A *challenge is to match the high-level business requirements with low-level affordances.* Specifically *real time* constraints related to energy management sub-systems with *high-level business concepts*. To that end we have suggested a *layered structure* of the SLAs as bundles of sub-SLAs. The high-level SLAs are *coordinated, via message exchanges*, with low-lever SLAs at the Information Layer of Figure 5. Some technological challenges are further discussed below (Section 5.3).

In present power systems the ICT system is mainly the SCADA system monitoring and controlling production and distribution of power. The SCADA system *integrates information from the component Domain with the Operation Zone for monitoring and control*. The operators react on information given to them with control signals down the zones. Since the use case at the business layer is a point – *a fixed contractö* between customer and distributor - we do not have to consider specific business information at the operator level. Use cases where we really need more complex SLAs are further discussed in Section 5.

4. ETRANSIT BUSINESS MODEL FRAMEWORK

From Section 2.4 we have seen that Smart grids can be modeled as *three* interconnected types of systems. The Energy management and Information management systems are the focus of SGAM frameworks Figures 4 – 6. To *complement* those views with a focus on *Business management* we suggest add to those these models the *eTransit Business Model Framework*. This framework was a result of the eTransit project in Sweden and Finland 2013-14 [10].

The focus of the eTransit project was to investigate the effects on the business processes of the rapidly expanding e-Commerce sector, specifically the role of mobile access points and its mutual dependencies with more traditional points of sale. A specific focus of the project was to investigate drivers and obstacle related to *transitions* between different business channels. To support those investigations we proposed the following eTransit Framework (Figure 8).

The eTransit Framework identifies a set, the *eTransit Views*, of useful aspects on Business Models (BMs). The Framework, with associated Methodology, is supporting *Requirements Engineering, Design, Implementations* and *Validation of* Business support systems. The *usefulness and adequacy* of the model was validated in the project as well. Since the Business management view of Smart grids, addressing energy based markets and empowerment of customers, is important, we suggest that the eTransit framework is a natural and necessary complement to the other two views of Smart grids given by SGAM (Section 2.4).

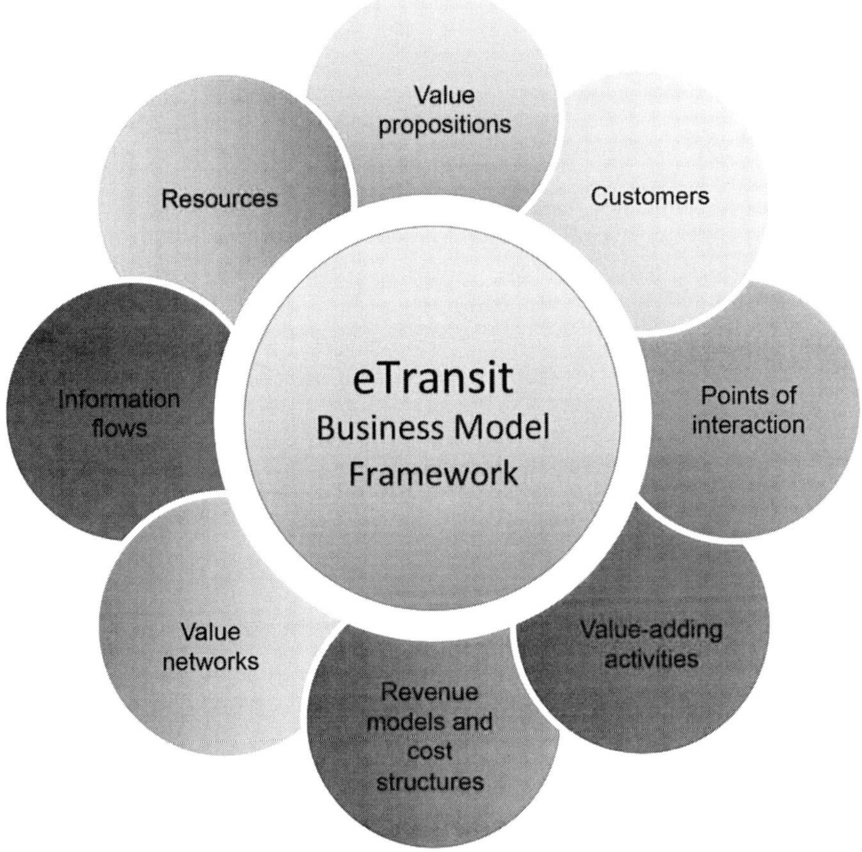

Figure 8. The eTransit Busness Model Framework

The Framework is an extension of BMs put forward by *Place to Space* [11], *Canvas*[6] - and *Visor*[7] - , to allow a deeper investigation and modeling of transitions between different *business channels, or implemented workflows. Vicwpoints* of the Framework are:

- Value Propositions
- Revenue models and cost structures
- Value network
- Value adding activities

- Information flows
- Resources
- Points of Interaction
- Customers

In our supplementary methodology we *configure* a selection of the eight different viewpoints into a solution of a given use case. This configuration constitutes *design and implementation of the corresponding SLA* of previous section.

The starting subset of eTransit supporting our specification of SLAs is (Section 3 and Section 5) is:

- *Value proposition* (VP): A package of a solution and a product / service offered to a specific customer segment of a use case
- *Revenue Model* (RM): Sources of revenue for a VP
- *Points of Interaction* (PI): Workflows related to a VP are implemented as different *channels* with information check points at PIs (Figure 9)
- *Resources* (Rs): Critical tools for a company in order to create and deliver its VP

The workflows are determined by the SLA of the use case. A workflow is a *chain of services* connected at PIs (Figure 9). The corresponding *information flow* has *inspection points* at PIs that are monitored and controlled by the SLA. Information related to *revenue models and cost structures,* of each workflow, is *collected for each stakeholder* of the SLA at the PIs.

We have suggested *separating the design and implementation of workflows in channels* (Figure 9), by using the identified BM models as *meta-models* for the channels. Specific *implementation models* for each channel are refinements of those meta-models meeting specific requirements for the selected channel (Figure 5 – 7). This allows *reuse and reconfigurations* of workflows in *different channels* with *invariant* meta-models (Figure 9). In short, it allows *cost efficient and reliable management and change of business channels.*

The starting points for design and implementation of SLAs are:

- *Identification from the use case* the related Value Proposition (VP), Revenue Models (RM), and Customer views (C)

- *Identification of* Value Network (VN) and Resources (R)
- *Identification of Information flows* (IFs), workflows (WFs) and Points of Interaction (PIs)

Key issues related to the translation of meta-models onto suitable implementation models include (Figure 6 and Figure 7):

- *Identification of suitable goal architectures* supporting flexible use
- *Identification of suitable sets of protocols* to support implementation, testing and validation
- *Agreements* on suitable *Key Performance Indicators* (KPIs) to be monitored
- *Selection* of suitable protocols and inspection methods at PIs to ensure *interoperability* and *trustworthiness* (Figure 9 and Figure 10)
- Selection of proper services to be trustworthy outsourced, e.g., payment services

The methodology has to be refined and tested but is based on similar methodologies supporting implementations of *Service Oriented Systems (*SOS) such as *Multi Agent Systems* (MAS) [6].

It is important to note that the identification of a VP, together with Revenue Models and Cost Structures, will provide a 'skeleton model' as a starting point. It should also be noted that *payment solutions* and the *roles and status of suppliers* are of big concerns for seller – buyers and should therefore be specifically addressed by the SLAs.

Related works in other domains are reported in [12]. In this case health care systems supporting patient oriented decision-making aiming at increasing patient safety.

4.1. Cognitive interaction patterns

The following Figure 9 illustrates a VP with associated WFs, implemented as channels, with PIs generating and controlling the related information flows. The information collected and processed will monitor and control the execution of a use cases in accordance with agreed upon SLAs. Related costs and revenues are collected and distributed among stakeholders.

Our findings from evaluations of the interviews in the project eTransit [10] suggest the following two complementary strategies to implement cost efficient and resilient e-Commerce solutions:

- Preparedness for flexibility and trustworthiness
- Minimizing costs for maintenance

The separation of meta-models and implementation models also allows assessments of *transitions* between different channels (for example mobile/stationary solutions) meeting different customer needs.

A closer analysis of different policies that can be implemented at PIs of different channels can greatly influence the decisions made by customers at those points.

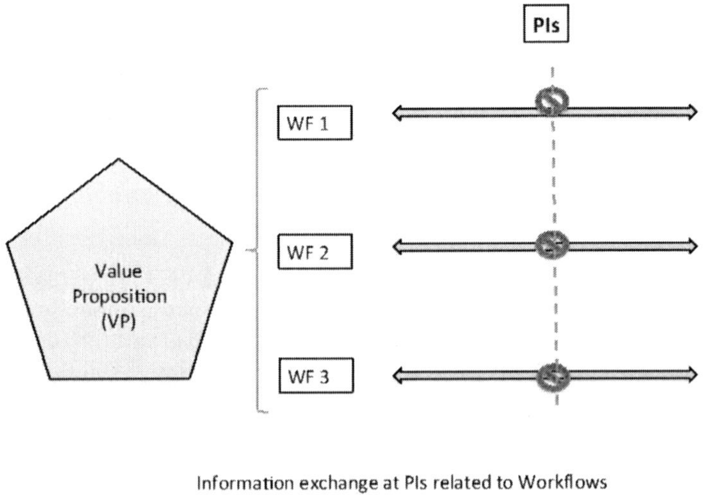

Figure 9. Some relations between VP, WFs and PIs

The purpose of our design and implementation of Cognitive Interaction Patterns is to translate a VP into Workflows to be assessed by potential customers. The goal is, of course to guide the customer to make an educated decision on the proposed VP! The next issue is thus related to decision support.

Classical Decision support systems are based on models of *Rational Reasoning.*

This subject has ben investigated from a computational point of view since 1950's. Research areas include *Artificial Intelligence* (AI), *Agent technologies* (AT) and *Multi-Agent Systems* (MAS). Rational reasoning is based on formal (mathematical) models such as different logics or statistical models. The strength is that we can reason about the reasoning (proofs or validations) [13, 14]. A weakness is that human decision-making is not always rational in the objective sense but often *subjective rational.*

There is at present high international focus on developing 'smart' distributed networked systems in areas such as Smart grids and Smart healthcare. Those kinds of 'smartness' can be modelled as service oriented systems (Multi-agent Systems) with intelligent interfaces (APIs) implemented using AI technologies [15].

Behavioural and experimental economists have in recent years successfully challenged the conventional view that individual actors make decisions purely by trading off expected costs and benefits (i.e., the "unbounded rationality" model) and suggested that behavioural predictions should be based on a *bounded rationality* framework instead [16]. The bounded rationality framework has generated a range of insights that apply to design and implementations of suitable PIs related to selected VPs!

By definition a *default* is a condition that is imposed when an individual fails to make a decision. Research and practice show that it is possible to *significantly impact people's behaviour* by carefully setting the default. For example, when asking for consent to store personal data online for marketing purposes, consent rates are higher when consent is the default (i.e., a "presumed consent" model) compared to a situation where the default is no consent (i.e., an *explicit consent* model) [17, 18, 19].

The bottom line is that designers of IP policies for e-Commerce applications should learn how to *customize* Interaction Patterns in a proper way to empower and learn more about their customers. In short we can combine rational information sharing between system components and bounded rational information sharing with or between human actors while maintaining interoperability assuring *shared and common understanding* of states of a use case [20, 21]. These insights are important in addressing Increasing Energy Efficiency by Reducing Losses and Promoting Value Propositions in future Smart grids.

5. SUITABLE COST AND REVENUE MODELS FOR FUTURE SMART GRIDS

Value Propositions and related workflows are illustrated in the following Figure 10. Inspection of message contents and distribution of messages to involved stakeholders of a VP is taken place at *Points of Interaction* (PIs) along the workflow. The workflow is implemented on suitable channels of the information systems.

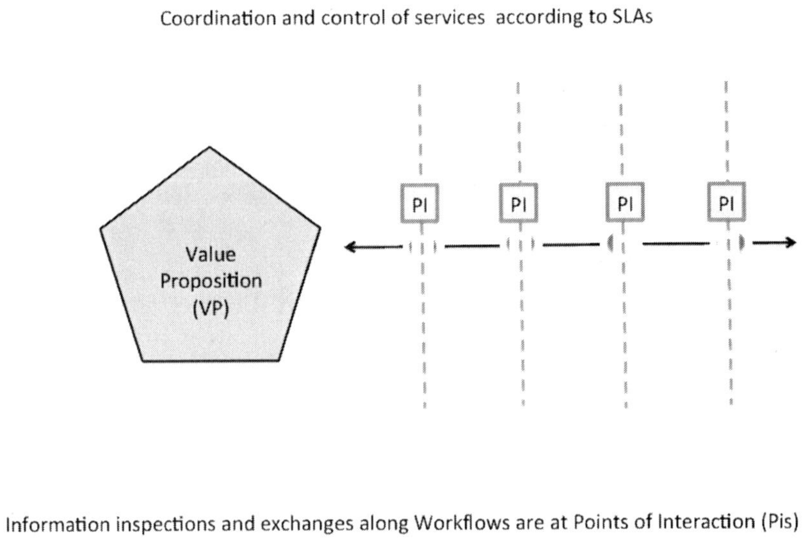

Figure 10. A VP of a use case involves stakeholders and services and information flows defined by the SLA.

The information monitoring and control of the workflow, governed by the SLA, is managed at the Points of Interaction (PIs) along the workflow (Figure 10). The infor-mation managementto all stakeholders of the VP is supported by suitable cognitive interaction patterns (Section 4.1).

There are different aspects of customer oriented Value Propositions. *Firstly*, we have VPs in application areas directly related to energy services. *Secondly*, we have VPs of services utilising parts of the Smart grid infrastructures to support, for example, comfort of living and/or health care.

In this chapter we focus on the following customer oriented VPs related to energy services:

- Power Quality (Section 5.1)
- Cutting losses in power distribution (Section 5.2)
- Smart Home Services enabled by Service Level Agreements (Section 5.3)

5.1. Power quality

In order to maintain the grid and to meet system performance requirements, the following set of ancillary services have been identified and implemented:

- Operational reserve
- Frequency control
- Reactive power and voltage control
- Short-circuit capacity
- Voltage quality
- Black-start
- Island operation of the grid
- Inertia
- Stability

The responsible stakeholders for quality and efficiency of the traditional regulated power systems are the utilities (TSOs and DSOs). Besides the main tasks of generation, transmission and distribution of power, the utilities also are responsible for *ancillary services related to system protection, Supervision and Control (SCADA), and billing*. Today, most of the ancillary support systems are hence *closed and proprietary*. In the traditional regulated power market those ancillary services were provided by the (state-own) utilities. The specific costs were implicit and as such hidden from the customers.

One of the main traits that differentiate existing power markets from the vertically integrated monopoly of the past is the *changing ownership of the product*, electric energy, as it moves from the generating plants (sources) to users premises. Although ownership changes hand with price variations, electrical energy is regarded as a homogenous product irrespective of which generating source it comes from and what resources are used to produce it. This is fortunately a sufficient definition for the power market to operate on as it is not necessary to have any differentiation when dealing with units of energy only.

However, this model is not suitable for handling power quality by itself. Power quality is classically not a sellable product. It is an attribute or value-added feature of another product, electrical energy. On the other hand, modern industrial processes with new types of electrical equipment are *increasingly adversely affected by poor power quality* than the traditional equipment. In addition integrated manufacturing supported by extensive automations and our today's individual way of life are more dependent on good quality of electricity than ever.

The goal of deregulation is to enhance competition and to bring to customers new choices and economic benefits. We argue that *Markets of Services based on power quality* could be based on *generalized* bilateral agreements between customers and providers. Those Service Level Agreements (SLAs) could become an *coordination and monitoring* mechanism between stakeholders of different energy based markets [27]. In the paper *A conceptual view of Power Quality Regulation Using Market-Driven Mechanism* [25] the authors suggest to employ market mechanisms to regulate power quality. However, in order to successfully identify and sell power quality services we have to identify and market value added Value Propositions (VPs) to customers. Those issues are treated in Section 4 and later in this section.

Challenges related to Ancillary services of Smart grids are mainly due to the distributed character of such systems involving new types of stakeholders and Distributed and Renewable resources (DER/RES) supporting active *prosumers (producers – consumers)* in agile and changing energy based business processes. The roles, stakeholders, responsibilities and business models of Smart grids are evidently completely different from those homogenous utility-based models of classical power grids. Focus of current R&D on Smart grids have been on energy-centric services and related smart energy markets. *Market models* related to *ancillary services* of Smart grids, e.g., business and cost models, remains to be identified and implemented. This will, of course, require changes of the corresponding regulatory frameworks. This necessary transition also requires deeper insights in the economic, societal, technical and environmental constraints to be assessed and resolved [23, 24].

Future smart grids will be based on two key ingredients: *intelligence* and *power electronics*. Intelligence or 'smartness' will be based on *agent technologies* applied to *components and interfaces* to allow context dependent con-figurations and instantiations of service based systems coordinated by Service Level Agreements [9].

Energy losses in the present power grid T&D are typically around 10%. Those losses are mainly due to resistance and imbalance of active and reactive power. The use of *FACTS* (Flexible AC Transmission Systems) has been successful for the Transmission net [22] and are now also investigated for the Distribution nets.

The importance of issues related to Power Quality (PQ) is mainly due to the increasing number of disturbing loads and increasing number of RES/DER. Main references are *Power Quality Indices in Liberalized markets* [23], *A Conceptual View of Power Quality Regulation Using Market Driven Mechanism* [25] and *Quality of Electricity Supply as a Service* [26].

We adopt the following definition of PQ [27]:

- The ability of power systems to operate loads without disturbing or damage them.
- The ability of loads to operate without disturbing or reducing the efficiency of the power system.

That is: PQ is a *common concern* between producers and consumers of power. That is, *suitable to address* as SLAs!

Power quality disturbances defined by *indices and objectives* are important tools in our demonstrators supporting Power quality Markets [23]. PQ indices represent a *compact way* to describe the characteristics of PQ disturbances as single numbers

An important class of PQ disturbances is due to voltage sags. Identifying and detecting voltage sags are therefore an important issue in identifying and monitoring SLAs between producers and consumers in a PQ Market. Proposed methods identifying sources of voltage sags (inferring responsibilities in cases of low probability of simultaneous disturbances) are based on, e.g., the following models [23]:

- Disturbance power and energy approach
- Slope of the system trajectory approach
- Real current component approach

- Resistance relay approach

As a case study we shortly describe the main features of the *Resistance sign* approach. The following Equivalent circuit for sag source detection is from [23].

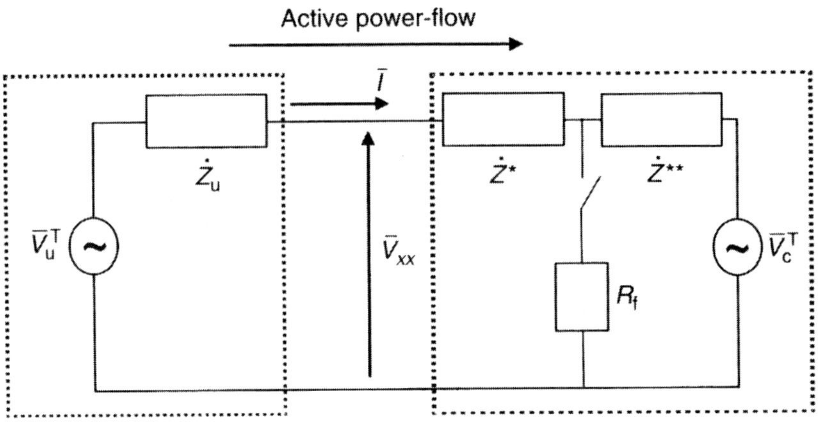

Figure 11. Equivalent circuit for voltage sag detection

Let us refer to Figure 11 in which the right side represents the customer installation ($Z^* + Z^{**} = Z_c$), while the left side represents the utility supply system. Setting up the equa-tions *at the customer side*, before and after a fault, and performing some calculations ([23] pp. 128 -129), we can deduce that:

- If Real (*Zc*) < 0, the impedance (multiplied by -1) is the *utility impedance* and the *fault is on the customer side.*
- If Real (*Zc*) > 0, the impedance is the *customer impedance* and the *fault is on the utility side.*
-

More examples of modelling and detecting responsibilities, related to Power Quality indices, between customer and utility are given in [24].

Economic aspects of PQ disturbances, based on economic schemes, have become increasingly important due to the current capability to quantifying the economic consequences associated with disturbances in power systems

[23]. The true economic value of PQ markets is linked to the effects that PQ disturbances have on equipment and other loads:

- Downtime disturbances causing lost production
- Additional costs related to maintaining equipment and reducing the effects of disturbances
- Existing methods for economic evaluation of PQ disturbances are mainly focused on voltage dips and harmonics

We argue that SLAs are useful tools to set up and maintain PQ markets. The PQ concerns between stakeholders in a use/business case can be expressed as SLAs using suitable indices and concerns. We can use our frameworks (Chapters 4 and 5) for setting up and monitoring suitable SLAs. In short:

- PQ agreements establish a *common concern* between producers and consumers of power. For example settling the investment costs on either side.
- Those concerns related to power disturbances can be expressed by selections of *suitable indices* and objectives
- The indices are composed by *measurable attributes* of the indices

The following Figure 12 captures the main ideas. The red rectangle illustrates the footprint of a SLA Coordinating services and stakeholders of a given use/business case (e.g., Power Quality markets). The footprint identifies interoperability Categories, layers of the GSAM framework (Figure 6), and selected Cross-cutting Issues identified by the SLA. The SLAs orchestrates Coordination, Monitoring and QoS of the services. Proper data aggregation, handling and transformations support semantic interoperability across the Smart Grid plane (Figure 5 and Figure 7).

Furthermore, during the different development phases the components of the SLA have been addressed. Specifically, issues related to *resolving potential conflicts* (investments) or interoperability issues.

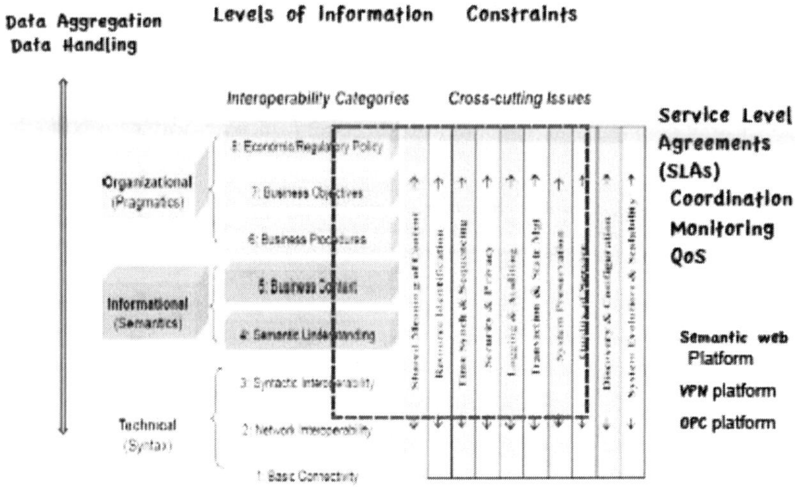

Figure 12. Footprint of SLA in the Interoperability framework

5.2. Cutting losses in power distribution grids

State estimation of the power system from real-time measurements is one of the most important elements of modern Energy Management Systems and are currently deployed and used by electric utility companies.

A promising technology to cut losses in power systems is to introduce *Phasor Measure-ment Units* (PMUs) to enhance efficiency, performance, and reliability in existing electric power system infrastructure [8] - . A *Phasor* refers to *measurable AC power components* of Figure 2. The PMUs are connected in *Wide Area Networks* (WAN) to allow for large area views of the network. Comparing data between points on an electric system is a good way to reveal stress on the system and home in on the source of the problem. PMUs *monitor the characteristics of electricity flowing through a particular location*, for instance, at the point where a generator connects to the bulk power system, or at a substation. The ability to compare *time-synchronized* data on the same timescale, among widely separated locations, is a relatively new achievement, based on:

- *Speed.* PMUs make measurements at short time intervals – typically 30 times per second – significantly faster than traditional SCADA systems

- *Synchronization.* All PMUs across an interconnection are kept in precise time synchronization using GPS. This synchronization provides the capability to easily compare systems data among geographically dispersed units, creating wide-area visibility across large power systems. This was not previously possible using older technology

The synchronized phasor measurement (PMU) technology is relatively new, and consequently several research groups around the world are actively developing applications of this technology. It seems clear that many of these applications can be conveniently grouped as follows:

- Power system monitoring
- Advanced network protection
- Advanced control schemas

The *North American SynchroPhasor Initiative* (NASPI) supervises instalments and assessments of several PMU installations [28].

A PMU collects and time-stamps data at a point. This data is then *processed and reacted on* according to a selected application.

In a *self-healing system* we have to implement the following cycle of tasks:

1. *Detection* of a fault
2. *Localisation* of the fault
3. *Repairing* of the fault
4. *Restart* of system

The PMUs can directly assist in the first two steps for grids. The remaining steps have to be addressed as add-ons to the PMU system.

We have designed and implemented a pilot of self-healing on Smart grids in the EU Project INTEGRAL [29].

5.2.1. Microgrids and renewable energy sources

We can think of the *Microgrid* as the great equalizer between consumers and the electrical companies. With centralized local power generation, businesses, communities and even towns can have a say in the generation and distribution of local energy.

A Microgrid is a scaled down version of the traditional power grid that most people are familiar with. A Microgrid can be as simple as having a power source like a generator hooked up to a load like a commercial business. A Microgrid can also consist of distributed energy resources (DER) like solar PV systems and wind turbines that have several electrical loads. These Microgrids can operate independently or in parallel with the traditional power grid.

Because Mirogrids can rely on local sources of power generation and also pose a lower demand on the grid infrastructure, large institutions like prisons, campuses and military operations, large commercial and industrial markets or remote settings that are off grid are realizing the benefits of building and maintaining their own Microgrid.

Benefits of Microgrids include:

- Increased efficiency – with the source of generated electricity so close to the use, very little energy is lost in transmission
- With fewer load sources, demand on the Microgrid infrastructure is less than a typical grid
- By being smaller and closer to source and demand and being able to use power generation more specific to the location, the system has higher reliability and is able to respond to demand more quickly
- Microgrids are laid out in a modular manner making expansion and updating more efficient
- With local control, both design and future planning are specific to the needs of the entities participating in the Microgrid
- Because the Microgrid can shut itself off from the main grid (islanding) it is less vulnerable to outside attacks, being cyber or physical

Barriers to the development and running Microgrids include:

- Lack of standards from operations, safety and integration to data on power quality from different sources.
- Legislation and regulations need to be addressed for regulating the operation of Microgrids in many countries.

- Installation costs of distributed energy resources such as solar and wind, may be too great for some areas.
- Lack of technical experience, infrastructure and communication protocols.
- Possibility of market monopoly if no infrastructure is in place to guard against pricing abuse.

Microgrids have to be fitted into the modernization of the current electrical grids that service the world. With increasing volatility to brown or black outs and physical or cyber attacks, Microgrids can keep important services running during any disruptions to the main grid.

We are presently testing technologies of Microgrids with basic components delivered by Hughes Power Systems [30]. We will implement a *Tests site* aiming at implementing smart protection and detection systems enabling, for instance, cutting losses in the distribution system. Specifically by installing products for minimising the number of interruptions (SAFI), the number of interruption minutes (SAIDI) and distribution losses.

We aim at implementing *self-healing capabilities of faults*. That is, *Smart Protection Systems for Microgrids*. Detection and Localisation is facilitated by new 3-Phase *Autoreclosers* (Vacuum Interupter Modules). Smart *automatic sectionalising* is enabled by Autoreclosers together with *distribution transformers* (with very low losses and with on-load tap). We will address a selected subset of the Technical losses listed in Section 2.3.2. Specifically losses related to:

- Lengthy Distribution lines
- Inadequate Size of Conductors of Distribution lines
- Installation of Distribution transformers away from loads centers
- Low Power Factor (PF) of primary and secondary distribution systems
- Bad Workmanship
- Feeder Phase Current and Load Balancing

The configuration of the Test site is given in Figure 13.

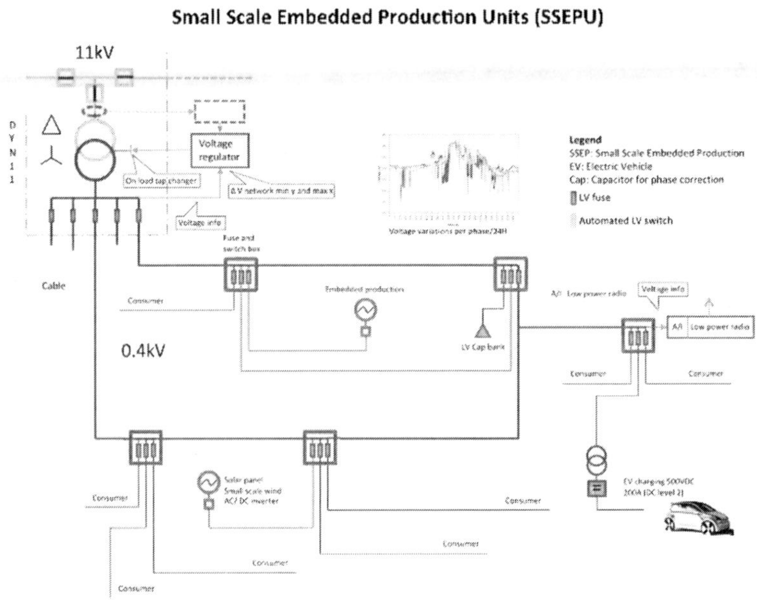

Figure 13. Smart equipment supporting cutting losses in distribution grids

The *Repair, Restoring and Reporting* services will be implemented as a set of suitable smart knowledge bases enabling collecting and handling the local information generated by the Autoreclosers.

Repairing might include reconfiguration of the distribution net to cut losses or increase performance. The reporting module also incorporates a learning mechanism to improve the total performance.

The experimental environment will also be a test bed for selected business cases related to energy based service markets.

5.3. Smart home services enabled by service level agreements

Smart and Green Building are active areas of R&D related to future Smart grids and active customers[9] - (Figure 14). We will select a use case illustrating *Comfort management* in Smart buildings, main own reference are [9, 31 and 32].

The concepts of Service Level Agreements (SLAs) have gain recognition and acceptance with the increasing demand of services in our societies. It is noticeable from ICT business offers, that tariffs are regarded as an old inflexible generic form of service level agreements. The deregulation of the telecommunication sector highlights the issues related to services and related technologies to meet the demands of new business opportunities. As a consequence, the organization *TeleManagement Forum* (TM Forum) has developed a set of technologies and tools, such as *Frameworks*, to support running and renewing service businesses in that sector.

Understandably, the work and results from TM Forum, including SLA management, are valuable inputs to design and implementation of SLAs for Smart grids. For instance, Smart grids presuppose a deregulation of the electric power market towards a service-oriented market, similar to de deregulation of the telecommunication market. However, there are some fundamental differences between the domains.

TM Forum work on SLAs is an important input to SLAs supporting Smart grids. How-ever, due to the different nature of smart grid we have to address the following specific additional challenges:

- SLAs between different groups of multiple stakeholders
- SLAs involving different sets of real-time constraints. That is, interactions between high- and low-level SLAs.
- SLAs supporting setting up and validating interoperability in Smart grids.

Firstly, formulation of agreements needs business cases with identified stakeholders (roles, capabilities and responsibilities). Secondly, relevant tasks to be coordinated among stakeholders should be identified. Thirdly, non-functional constraints (perfor-mance, security, Quality of Service, etc.) should be identified and taken into account. To do that, the stakeholders should agree upon KPI (Key Performance Indicators) to be implemented and monitored according to the agreements.

Important application areas of Smart Grids are Smart and Green Buildings, Figure 14.

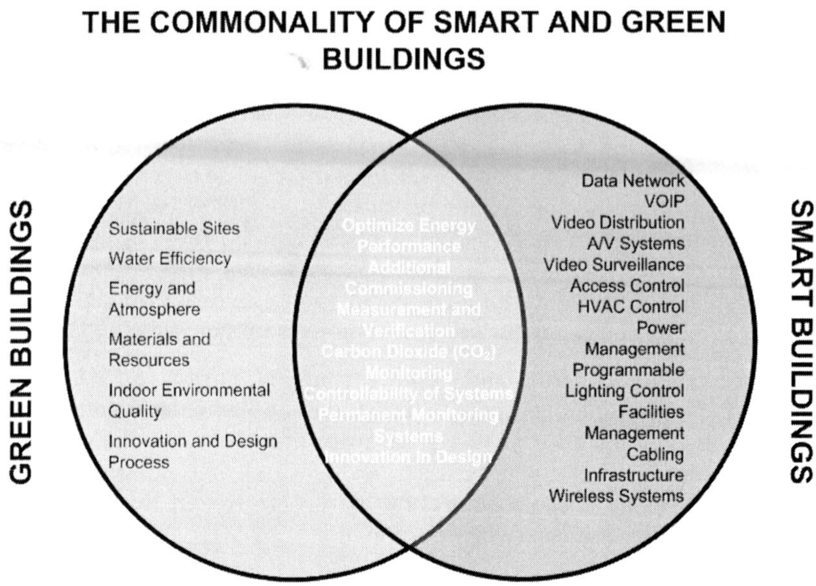

Figure 14. Common application areas of Smart and Green Buildings

Our selected use case, smart home services, is a typical example where we need coordination between high-level and low-level SLAs (Figure 7). The high-level SLA guarantees that the *user perceives and accepts* the desired comfort level, set by the control system. The low-level SLA monitors and *controls the settings of the associated physical system*. Thus we have to address and solve the *challenge of trustworthy translation between perceived subjective parameters* and *settings of physical objective parameters*.

As presented in Figure 7, the key drivers in more complex business cases of Smart grids are SLAs that holds a strong interaction between the higher-level business cases and low-level affordances of the infrastructure.

We model our use case as follows:

Stakeholders

- Customer (C)
- Smart house agent (SHA)
- Market aggregator (MA)

- Operation agent (DSO)
- Service provider (SP)

Scenario

The Market Aggregator (MA) offers a range of services (VPs Figure 8) to the Customer (C). Among those is Comfort management. This service is delivered to C by the SHA in cooperation with SP. SP is responsible for instalment, monitoring and control of smart equipment at customer premises. DSO, together with SHA and MA monitor and control the energy flow associated with a given service. The MA is also responsible for the monitoring and billing of the services to C.

Implementation

The Comfort management systems is modelled and implemented as a *Multi Agent System* (MAS) consisting of the agents C, MA, SHA, DSO and SP. The coordination and control of the MAS is governed by the factual set of High-level and Low-level SLAs. Agents are grouped into the two bundles of SLAs. Firstly, a high-level business related set, SLA-high, between the agents C, SHA, SP, and MA. Secondly, a real-time constrained set represented as SLA-low, between agents MA, DSO, SP, and SHA. SP runs coordination between the SLA sets.

Challenges

The comfort level identified by the customer is represented in a *set* {good, acceptable, not acceptable, not-good}. The following physical (objective) parameters of the customer premises are monitored and controlled by SP:

- Temperature
- Lighting
- Sound
- Humidity
- Air circulation
- Allergy indicators
- Movement indicators

A design challenge is to *identify, implement and validate* an *accepted translation* between the objective sensor readings and the subjective customer values (Figure 5 and Figure 7). This implemented mechanism provides *trustworthy interoperability* of meaning and fulfilment of services according to the SLA (Figure 6).

Requirement Engineering and Validation

Negotiation of a SLA provides a methodology supporting requirement engineering of SLAs. SLA negotiation tools provides a mechanism to *identify* the concerns of the involved stakeholders in a given use case. Typically, those concerns are expressed in *stakeholder case specific concepts* (ontologies). The next step is to negotiate a common understanding among stakeholders of related concepts.

There are in principle two ways to discover and validate this pattern-matching; *model based* or *experimental based*.

A model-based approach is based on establishing a *formal function* between the different types of parameters. That is:

Comfort is defined by an analytic function F:

F(objective_parameters) with value-set {Good, Acceptable, Not acceptable, Not good}

However, this approach is *neither easy nor flexible*. In an *experimental setting* we provide a flexible front-end to the customer with the SP operating the sensor net at the premises of C. By performing a rich enough *calibration* between sets of objective parameters and the customer perceived comfort level we derive a set of tuples Σ, or patterns, connecting objective parameter settings and subjective comfort levels.

Σ(objective parameter values, comfort levels)

Proper analysis of this set will reveal *sufficient rich patterns to support design of corresponding SLA rule sets.* Setting up the front-end to the customer can *support validation* of the rule sets (Figure 6). Maintenance of the rule sets could also be supported with a similar set up.

Management of SLAs

After calibration and validation, the system build substantial knowledgebase that can facilitate decision-making based on historic data of the user's behaviour further supporting negotiations of the corresponding SLAs. The customer is given a set of choices to select the desired type of SLA related to comfort levels and price. That is, the *Value Proposition*! After the negotiation, the system provides monitoring mechanism of SLA to ensure proper QoS of the service (Figure 10).

Exception Handling: In case of indication of system comfort output of "Not acceptable" or "Not good", generate alerts. The same holds if there is detection of malfunctions of system components.

6. CONCLUSIONS AND FUTURE WORK

This chapter is about Increasing Energy Efficiency by *Reducing Losses* and *Promoting Value Propositions* in AC power systems. *Reducing losses* relates to the *physical processes* of generating, transmission and distribution of electric power. *Promoting Value Propositions* relates to identifying customer oriented value added business cases and processes based on energy based services.

The core of the emerging Smart grid [5] is addressing the challenges of balancing a very complex physical system in real time while meeting customers and societies expectations on energy and energy related reliable services. The overall constraint, however, is meeting the environmental challenges!

In order to give perspectives on those societal challenges we have given some technological as well as business and societal backgrounds and illustrated our suggested solutions with some implemented examples.

Future Smart grids will consist of the following three interconnected types of systems:

1. Systems supporting *Energy management*
2. Systems supporting *Information management*
3. Systems supporting *Business management*

The technological background of systems of types I and 2 are grounded in *Smart Grid Reference Architecture* (SGAM) Figure 4, Figure 5 and Figure 6 [4]. Those architecture models illustrate the *inherent complexity* of designing, implementing and maintaining Smart grid solutions.

To complement the views of the SGAM architectures of system 1 and 2 we suggest an *extension* of the SGAM model sets with the *eTransit Business Model framework* (Section 4). This enables modelling the views of System 3 and but also the *interactions* between all three types of system (Section 5).

To deal with this complexity, in *time, space and cyberspace*, we propose *Service Level Agreements* (SLAs) as a *coordinating and monitoring mechanism of business cases*. Figure 7 illustrate how bundles of SLAs can manage the different aspects and types of business use cases.

SLAs can be properly identified and modelled by a vertical and horizontal mapping of a use case onto the SGAM (Section 3). To proper model business cases from different viewpoints we propose the eTransit Business Model, Figure 8. This framework allows for implementing SLAs into Workflows, Channels and Information flows (Figure 9 and Figure 10).

In this chapter we focus on models supporting *trustworthy information exchange between stakeholders in a business case*. In coordination of a use case interoperability is the *shared understanding by stakeholders of the state of the executed use case.* This requirement is extremely important [20, 21]. A key technology is here designing and implementing *Cognitive Interaction Patterns* (Section 4.1)

Societal demands on Smart grids include requirements on *inclusion of Renewable Energy Resources (RES*) and efforts to decrease the environmental negative effects of energy production. To illustrate some of the challenges we indicate the expected total energy consumption 1990-2040 (Section 2.1). *Challenges and problems in comparing* different energy sources are discussed in Section 2.2.

In Section 5 we give some examples of customer centric use cases (Value Propositions) to increase energy efficiency and/or to add customer value:

- *Power Quality* (Section 5.1)
- *Cutting losses in power distribution* (Section 5.2)
- *Smart home Services* (Section 5.3)

Future work includes investigations of *Value Propositions* (VPs) in areas of comfort or Health. Those VPs can be based on energy systems such as Microgrids (Section 5.2.1 and Section 5.3) as in Figure 14.

In defining and discussing proper *Cost and Revenue Models* we refer to Figure 1 with the interpretations; *Total Cost, Customer paid cost* (Value Proposition) and *Investment cost* (Private & Public) giving a balance that drives societal acceptable products and services.

Finally, it should be observed that future Smart grids have some commonalities and differences with systems classified as instances of *Cloud computing*[10] - · In this setting Smart grids could be modelled *three interacting Platform –as – a – Service* (*PaaS*) cloud systems. That is, as we earlier have stated: Energy system management, Information management systems and Business management systems. Having said that, we believe that Smart grids and Cloud computing systems could benefit from sharing lessons learned.

This observation is enforced by the *Report on the public consultation for the EU H2020 Work Programme 2016-17: Cloud Computing and Software*, Final version 2014-12-04[11] - , that states:

"Attention should be placed on the non-functional aspects of cloud-based applications and services including SLAs and performance optimisation of real-time, cloud based applications and services, and increased energy efficiency through the exploitation of hardware advances, lightweight virtualisation techniques, and intelligent auto-scaling algorithms."

7. ACKNOWLEDGEMENTS

We gratefully acknowledge the cooperation in the eTransit project with our colleagues from the School of Business and Economics at Linneaus University in Sweden and Institute for Advanced Management Systems Research (IAMSR) at Åbo Akademi University in Finland with the support from Peter Wallenbergs Foundation.

NOTES

[1] - http://en.wikipedia.org/wiki/AC_power

[2] - *http://www.edf.org/climate/how-cap-and-trade-works*

[3] - http://en.wikipedia.org/wiki/Cost_of_electricity_by_source

[4] - *http://en.wikipedia.org/wiki/Electric_power_transmission*

[5] - http://electrical-engineering portal.com/

[6] - *http://www.businessmodelgeneration.com/canvas*

[7] - *http://classic.marshall.usc.edu/ctm/research/visor-business-model.htm*

[8] - http://en.wikipedia.org/wiki/Phasor_measurement_unit

[9] - http://ec.europa.eu/information_society/activities/sustainable_growth/docs/sb_publications/smartbuildings-ld.pdf

[10] - *http://en.wikipedia.org/wiki/Cloud_computing*

[11] - EGI recommendations for the Horizon 2020 Work Programme 2016-2017, https://documents.egi.eu/document/2320

REFERENCES

1. U.S. Energy Information Administration (2013): *International Energy Outlook 2013.* DOE/EIA 0484(2013)
2. Frank, C. R (2014): *The Net Benefits of Low and No-Carbon Electricity Technologies.* Global Economy and Development at Brookings. Working Paper 73, May 2014.
3. Parmar, J (2013): Total losses in Power Distribution and Transmission Lines. In *Transmission and Distribution Electrical Engineering Portal (EEP).* Posted August 19, 2013
4. European Smart GridsTechnology Platform: Home page http://cordis.europa.eu/technology-platforms/smartgrids_en.html
5. European Technology Platform (2012): *SmartGrids SRA 2035* – Strategic Research Agenda. Update of the SmartGrids SRA 2007 for the needs by the year of 2035
6. Gustavsson, R. and Hussain, S. (2014): The Proper Role of Agents in Future *Multi-Agent Systems PAAMS 2014 International Workshops*, pp. 226 – 237. Springer Verlag.
7. Gustavsson, R. (2012): Promoting Increased Energy Efficiency in Smart Grids by Empowerment of Customers. Invited book chapter in *Modeling and Optimization of Renewable Energy Systems.* InTech Open Access Publisher, 2012, ISBN 7978-953-308-487-9
8. CEN-CENELEC-ETSI Smart Grid Coordination Group (2012): Smart Grid Reference Architecture (SGAM), v 3.0. November 2012
9. Gustavsson, R., Hussain, S. and Nordström, L (2011): Engineering of Trustworthy Smart Grids Implementing Service Level Agreements. In *Proceedings of 16*th *Conference on Intelligent System* Applications to Power Systems. September 25 – 28, 2011, Greece, 2011
10. The eTransit project (2013): A collaboration project between School of Business and Economics Linneaus University in Sweden and Institute for Advanced Management Systems Research (IAMSR) Åbo Akademi University in Finland with a grant from Peter Wallenbergs Foundation

11. Weill, P. and Vitale, M. (2001): *Place to Space. Migrating to eBusiness Nodels.* Harvard Business School Press; Boston, Massachusetts

12. Ådahl, K. and Gustavsson, R. (2011): Decision Support by Visual Incidence. Invited book chapter in *Efficient Decision Support Systems: Practice and Challenges – From Current to Future.* InTech Open Access Publisher

13. Shapiro, S.C (Editor) (1992): *Encyclopedia of Artificial Intelligence.* John Wiley & Sons, Inc.

14. Pearl, J. (2000): *Causality Models, Reasoning, and Inferences.* Cambridge University Press.

15. Wooldridge, M. (2002): *An introduction to MultiAgent Systems.* John Wiley & Sons Ltd.

16. Kahneman, D., (2011): Thinking Fast and Slow. Pinguin Books.

17. Brown, C., and Krishna, A. (2004): The skeptical shopper: A metacognitive account for the effects of default options on choice. *Journal of Consumer Research*, 31, 529-539.

18. Herrmann, A., Goldstein, D.G., Stadler R., Landwehr, J.R., Heitman, M., Hofstetter, R., and Huber, F. (2011): He effect of default options on choice – Evidence from online product configuration. *Journal of Retailing and Consumer Services*, 18, 483-491.

19. Menzel, S. (2013): Are emotions to blame? – The impact of non-analytic information processing on decision-making and implications for fostering sustainability. *Ecological Economics*, 96, 71-78

20. Lundberg, J. and Gustavsson, R. (2009): Robust approach toward context dependent information sharing in distributed environments. In *Proceedings of 11*th *International Conference on Enterprise Information Systems,* Milan, Italy, 2009

21. Lundberg, J. and Gustavsson, R. (2011): Challenges and opportunities of sensor based user empowerment. In *Proceedings of 8*th *International Conference on Networking, Sensing and Control ICSNSC 2011*, Delft, The Netherlands, 2011

22. P.K Steimer, P.K. (2011): Enabled by High Power Electronics – Energy efficiency, Renewables and Smart Grids. *Proceedings* of *The 2011 International Power Electronics Conference.*

23. Pierluigi, C., Carpinelli, G., and Verde, P. (2009): *Power Quality Indices in Liberalized Markets.* Wiley 2009. ISBN: 978-0-47003395-1

24. Caramia, P., Carpinelli, G., and Verde, P. (2009): Assessing Responsibilities between Customer and Utility. Book Chapter in *Power Quality Indices in Liberalized Markets.* Wiley 2009.

25. Chen, S., Wang, C.J., and Lie, L.T. (2006): A Conceptual View of Power Quality Regulation Using Market Driven Mechanism. *Proceedings of PowerCon 2006.*

26. E. Sersen, E. Vorsic, J. (2008): Quality of Electricity Supply as a Service. In *Proceeding of International Conference on Renewable Energy and Power Quality (ICREPQ-08),* 2008

27. Gustavsson, R., Hussain, S. and Saleem, A. (2013): Ancillary Services for Smart Grids – Power Quality Markets. In *Proceedings of PowerTech 2013,* Grenoble

28. NASPI homepage: https://www.naspi.org/

29. INTEGRAL Report: http://www.integral-eu.com/fileadmin/user_upload/downloads/Presentations/E-Energy2010_-_Integral__Oct-2010_.pdf

30. Hughes Power Systems. Home page: http://www. Hughespowersystem.com/en/

31. Hussain, S., Gustavsson, R., and Nordström, L. (2012): Smart Home Services Enabled by Service Level Agreements. In *Proceedings of First International Workshop on Intelligent Agent Technology. Power Systems and Energy Markets (IATEM 2012).* Vienna 4th September 2012.

32. Gustavsson, R. and Hussain, S: (2014): The Proper Role of Agents in Future Resilient Smart Grids. In *Proceedings of Multi-agent Applications for Smart Grids and Sustainable* Energy Systems (MASGES 14). Salamanca 4 – 6 June 2014.

7

CHAPTER

TOWARDS IMPLEMENTATION OF SMART GRID: AN UPDATED REVIEW ON ELECTRICAL ENERGY STORAGE SYSTEMS

Md Multan Biswas[1*], Md Shafiul Azim[2], Tonmoy Kumar Saha[2], Umama Zobayer[3], Monalisa Chowdhury Urmi[3]

[1]*Department of Electrical and Electronic Engineering, Bangladesh University of Engineering and Technology, Dhaka, Bangladesh;*

[2]*Department of Electrical and Electronic Engineering, Khulna University of Engineering and Technology, Khulna, Bangladesh;*

[3]*Department of Electrical and Electronic Engineering, Stamford University Bangladesh, Dhaka, Bangladesh.*

ABSTRACT

A smart grid will require, to greater or lesser degrees, advanced tools for planning and operation, broadly accepted communications platforms, smart sensors and controls, and real-time pricing. The smart grid has been described as something of an ecosystem with constantly communication, proactive, and virtually self-aware. The use of smart grid has a lot of economical and environmental advantages; however it has a downside of instability and unpredictability introduced by distributed generation (DG) from renewable energy into the public electric systems. Variable energies such as solar and wind power have a lack of stability and to avoid short-term fluctuations in power supplied to the grid, a local storage subsystem could be used to provide higher quality and stability in the fed energy. Energy storage systems (ESSs) would be a facilitator of smart grid deployment and a "small amount" of storage would have a "great impact" on the future power grid. The smart grid, with its various superior communications and control features, would make it possible to integrate the potential application of widely dispersed battery storage systems as well other ESSs. This work deals with a detailed updated review on available ESSs applications in future smart power grids. It also highlights latest projects carried out on different ESSs throughout all around the world.

Keywords: Battery; Distributed Generation; Hybrid Energy Storage Systems; Power Quality; Smart Grid

1. INTRODUCTION

The world's electricity systems face a number of challenges, including aging infrastructure, continuous increase in demand, the integration of growing numbers of variable renewable energy DGs and electric and hybrid electric vehicles, the need to improve the security of power supply, and the need to lower carbon dioxide (CO_2) emissions [1-3]. These challenges must be addressed also with regard to each region's unique technical, economic, and commercial regulatory environment [4]. Smart grid technologies offer ways not just to meet these challenges but also to develop a sustainable energy supply that is more energy efficient and more affordable.

Compared to other industries, our electrical grid has been largely bypassed by technological innovation until relatively recently, owing to the fact that historically it has been heavily regulated and modeled to keep the lights on and costs low. Partly for this reason, its modernization by means of information-technology tools and techniques has been somewhat of a back-burner priority. Like the telecom and internet revolutions that preceded it, technology holds the key to the smart grid and its realization. The smart grid and the technologies embodied within it are an essential set of investments that will help bring our electric grid into the 21st century using megabytes of data to move megawatts of electricity more efficiently, reliably, and affordably [5-9].

The smart grid will:

- Provide power quality for the digital economy;
- Accommodate all generation and storage options;
- Enable new products, services, and markets;
- Optimize asset utilization and operate efficiently;
- Enable active participation by consumers;
- Anticipate and respond to system disturbances;
- Operate resiliently against attack and natural disaster.

Storage is perhaps the most important smart grid advanced component because of its key role in complementing renewable generation. With the proper amount and type of storage broadly deployed and optimally controlled, renewable generation can be transformed from an energy source into a dispatchable generation source [10-12]. And with the addition of energy storage, more wind and solar generation can be added to a typical power system that employs a large percentage of slow-response fossil and nuclear generation. It is feasible that, the penetration of renewables can be significantly above 20 percent with the addition of sufficient energy storage technologies [13].

This paper is organized as follows. Section 2 presents a brief description of smart grid technology with its essential features. The necessity and prospect of energy storage systems in future smart power grid are broadly discussed in Section 3. Finally, Section 4 elaborately reviews the available energy storage technologies that are considered for use in smart grid applications with updated technologies, which is followed by some concluding remarks in Section 5.

2. SMART GRID: TECHNOLOGY DESCRIPTION

Though there has been much debate over the exact definition of smart grid, it actually comprises a broad range of technology solutions that optimize the energy value chain. Depending on where and how a specific utility operates across that chain, it can benefit from deploying certain parts of a smart grid solution set. Smart grid is a large electricity network that uses digital and other advanced technologies to improve efficiency, reliability, and security of the electric system: from large generation, through the delivery systems to electricity consumers and a growing number of distributed-generation and energy storage resources [14-16]. Smart grids co-ordinate the needs and capabilities of all power generators, grid operators, end-users, and electricity market stakeholders to operate all parts of the system as efficiently as possible, minimising costs and environmental impacts while maximising system stability, reliability, and resilience. Smart grids are an evolving set of technologies that will be deployed at different rates in a variety of settings around the world, depending on local commercial attractiveness, compatibility with existing technologies, regulatory developments, and investment frameworks. Figure 1 demonstrates the evolutionary character of smart power network [4].

Advanced metering infrastructure (AMI) is an approach to integrating consumers based upon the development of open standards. It provides consumers with the ability to use electricity more efficiently and provides utilities with the ability to detect problems on their systems and operate them more effectively [17,18]. AMI enables consumer-friendly efficiency concepts like "Prices to Devices" to work like this: Assuming that energy is priced on what it costs in near real-time—price signals are relayed to "smart" home controllers or end-consumer devices like thermostats, washer/dryers, and refrigerators—the home's major energy-consumers [19-21]. The devices, in turn, process the information based on consumers' learned wishes and power accordingly. The house or office responds to the occupants, rather than vice-versa. Because this interaction occurs largely in the background with minimal human intervention, there's a dramatic savings on energy that would otherwise be consumed [14]. Far more than "smart meters," a fully-functioning smart grid will feature sensors throughout the transmission and distribution grid to collect data, realtime two-way communications to move that data between utilities and consumers, and the computing power necessary to make that intelligence actionable and transactive as shown in Figure 2. Indeed, only by bringing the tools, techniques, and technologies that enabled the internet to the utility and the electric grid is such a transformation possible [8].

Smart Grid's Principal Characteristics

A smart power grid brings the power of networked, interactive technologies into an electricity system, giving utilities and consumer's unprecedented control over energy use, improving power grid operations, and ultimately reducing costs to consumers. In brief, the main features of smart grids are [9,22-26]:

1) A smart grid accommodates not only large, centralised power plants, but also the growing array of customer-sited distributed energy resources.

2) Not all commercial enterprises, and certainly not all residential customers, need the same quality of power. A smart grid supplies varying grades (and prices) of power.

3) Correctly designed and operated markets efficiently create an opportunity for consumers to choose among competing services.

4) A smart grid applies the latest technologies to optimize the use of its assets. For example, optimized capacity can be attainable with dynamic ratings, which allow assets to be used at greater loads by continuously sensing and rating their capacities.

5) Provides resiliency to disturbances, attacks, and natural disasters. Self-healing actions result in reduced interruption of service to consumers and help service providers better manage the delivery infrastructure.

6) Consumers help balance supply and demand, and ensure reliability by modifying the way they use and purchase electricity.

3. ENERGY STORAGE—A KEY ENABLER OF SMART GRID

Liberalization of the power market and widespread use of distributed energy resources (DERs), in particular DG and ESSs, could enable smart grids to have a significant influence on electricity market prices and ancillary services [27]. As focused in [27], the charge/discharge of various ESSs can be controlled in order to guarantying proper applications. Further, in modern power distribution systems, where a significant amount of the total electricity demand is met by renewable generation, ESSs can mitigate the uncertainties of energy sources (such as solar and wind) and can store the energy during high renewable production and/or low price periods, and deliver when either necessary or convenient [28-31]. Based on the ESSs technologies, in [27] the applications of ESSs are classified in instantaneous, short-, midand long-term. Instantaneous and short-term applications are involved in real time regulations, for example aiming at ancillary services provision or integration of electric drive vehicles batteries in the networks [32,33].

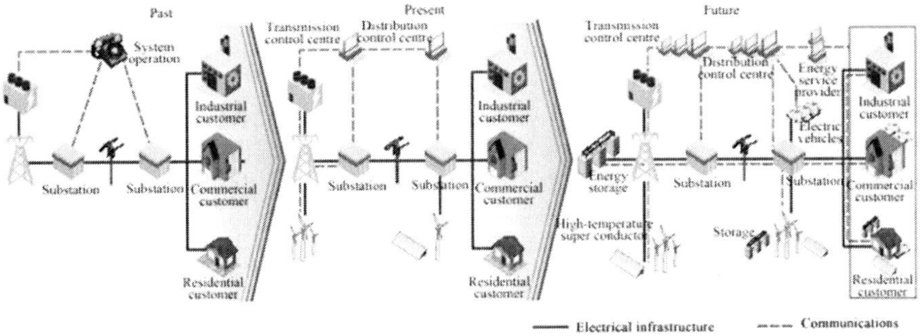

Figure 1. Smarter electricity systems.

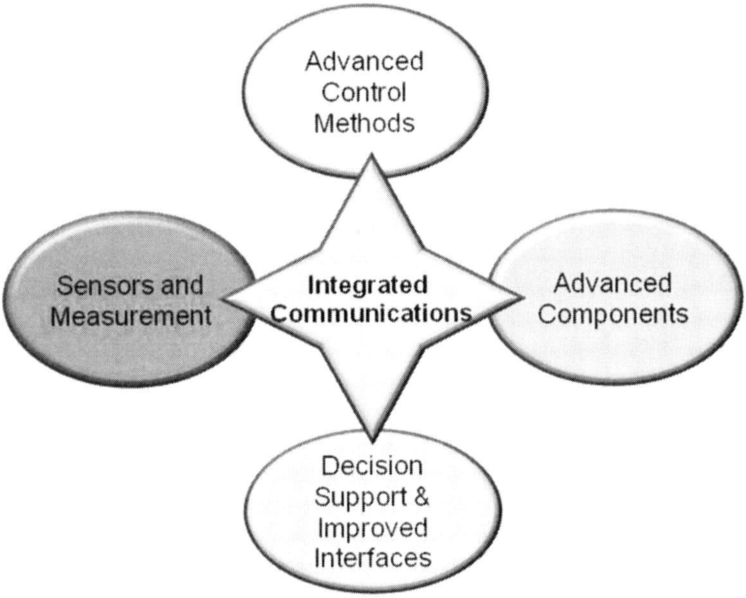

Figure 2. Advanced communication and control in smart grids.

Energy storage systems improve the efficiency and reliability of the electric supply system by reducing the requirements for spinning reserves to meet peak power demands [34]. Storage devices can also provide frequency regulation to maintain the balance between the network's load and power generated, and they can achieve a more reliable power supply for high tech industrial facilities as shown in Figure 3.

Figure 3. Distributed energy storage in a Windfarm.

High voltage power electronics, such as switches, rectifiers, inverters, and controllers, allow electric power to be precisely and rapidly controlled to support long distance transmission [35,36]. Thus, energy storage and power electronics hold substantial promise for transforming the electric power industry [28]. This capability will allow the system to respond effectively to disturbances and operate more efficiently, thereby reducing the need for additional infrastructure.

4. AVAILABLE ENERGY STORAGE SYSTEMS

The adoption of smart power grid devices throughout utility networks will effect tremendous change in grid operations and usage of electricity over the next two decades. Increased deployment of energy storage devices in the distribution grid will help make this process happen more effectively and improve system performance. The energy storage systems (ESSs) applicable in power networks can be divided into two major categories [29] as shown in Figure 4. The first category being largescale storage systems that can be used in utility transmission applications. The second group includes small-scale storage systems sited at the consumer's premises. Some of the available energy storage systems are potentially discussed in the following sub-sections.

4.1. Battery

Battery is one or more electrochemical cells that convert chemical energy into electrical energy. Batteries have been using for energy storage purpose for over one-hundred years and possess some very important, unique, and desirable features. Battery energy storage systems (BESSs) are modular, quiet, and non-polluting [37]. Batteries are manufactured in a wide variety of capacities ranging from less than 100 watts to modular configurations of several megawatts. As a result, batteries can be used for various utility applications in the areas of generation, transmission and distribution, and customer service and they can be installed relatively quickly. Battery has convenient size and voltage characteristics. The operating principle of a typical battery is shown in Figure 5.

A grid-scale BESS consists of a battery bank, control system, power electronics interfacing circuit for ac-dc power conversion, protective circuitry, and a transformer to convert the BESS output to the desired transmission or distribution system voltage level [38]. The battery bank consists of numerous batteries connected in a combination series-parallel configuration to provide the desired energy and power capabilities for the application.

The use of Lead Acid batteries for energy storage dates back to mid-1800s. There have been developed several new battery technologies to store more energy, last longer and less cost than the Lead Acid battery. Battery technologies differ widely in terms of their energy and power densities, energy efficiencies, cycle-life, availability, and operating conditions. Some of these new battery technologies are Lithium Ion, Lithium Polymer, Nickel Metal Hydride (Ni-MH), Vanadium Redox (VRB), Nickel Cadmium (Ni-Cd), Sodium Sulfur (NaS), and Zinc Bromide [39-41]. Table 1 summarizes the characteristic parameters of different batteries [27,28, 42- 44].

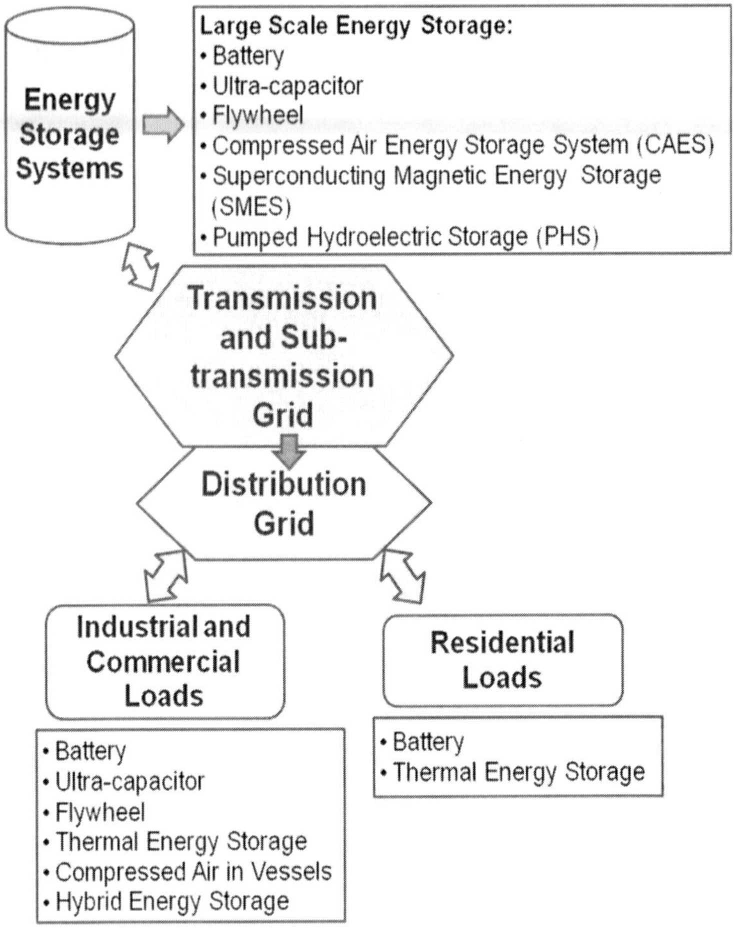

Figure 4. Energy storage systems in smart power networks.

4.2. Ultra-Capacitors

Ultra-capacitors (UCs), also known as super-capacitors, or electric double-layer capacitors (EDLCs), store energy in the electrical double layer at an electrode/electrolyte interface. There are no chemical reactions involved in the UC's energy storage mechanism. The main parts of an electrochemical capacitor are electrodes and electrolyte. The electrical energy E accumulated in ultracapacitors is related to the capacitance C or the stored charges Q and voltage V by following formula:

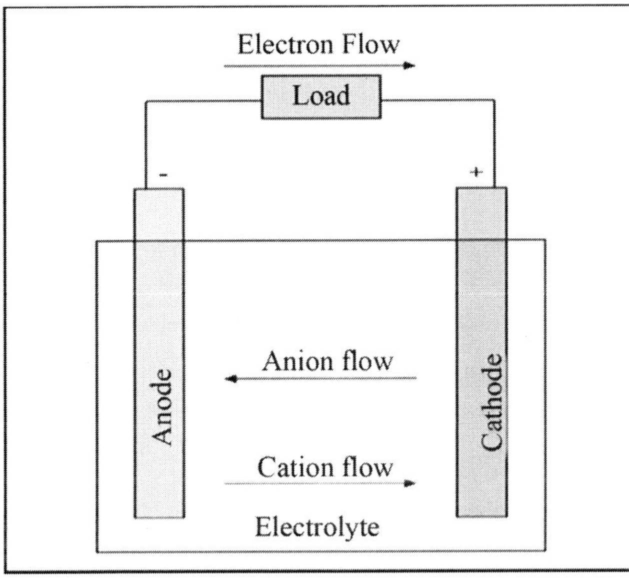

Figure 5. Working principle of a battery.

Table 1. Comparison of different types of batteries.

Type	Energy Density (Wh/kg)	Energy Efficiency (%)	Power Density (W/Kg)	Cycle Life (Cycles)	Self Discharge (%/Month)
Lead-Acid	30 - 40	70 - 90	180	200 - 2000	3 - 4
Li-Ion	100 - 250	75 - 90	1800	500 - 2000	5 - 10
Li Polymer	130 - 200	70	3000	>1200	4 - 8
Ni-MH	30 - 80	70	250 - 1000	500 - 100	30
Ni-Cd	40 - 60	60 - 90	140 - 180	500 - 2000	10 - 15
NaS	150	80 - 90	120 - 150	2500	-
VRB	25 - 40	80	100 - 150	>16,000	<1
Zinc Bromide	70	70	-	1000	-

$$E = \frac{1}{2}CV^2 = \frac{1}{2}QV \tag{1}$$

And,

$$Q = CV = \frac{A\varepsilon}{d}V \tag{2}$$

As for the conventional capacitor, the capacitance C is proportional to the area A of the plates and the permitivity of the dielectric ε and is inversely proportional to the distance d between the plates. UCs are designed to have a very high electrode surface area and use high permitivity dielectric. Due to the high permeability and close proximity of the electrodes, UCs have a low voltage withstand capability (usually 2 - 3 V) [30]. The electrode surface area is maximized by using porous carbon as the current collector, allowing a relatively large amount of energy to be stored at the collector surface. Therefore, UCs attain very high capacitance ratings. Larger UCs have capacities up to 5000 farads [45].

Ultracapacitors store energy by physically separating unlike charges. UCs have a long cycle life due to the fact that there are no chemical changes on the electrodes ideally in normal operation and UCs have superior efficiency. UCs also provide exceptional power density, since the charges are physically stored on the electrodes. Conversely, energy density is low since the electrons are not bound by chemical reactions [30]. The UC is temperature resistant with an operating range between −40°C to +65°C and is also shock and vibration resistant [46, 47].

In order to be a viable alternative in a large scale energy storage system they will need to be able to handle multiple kV. The ability of modular, non-polluting, quiet, quick charge and discharge capability, long life (10 to 12 years), and very high cycle life makes the UC a very desirable energy storage device. They can be use for short term ride through capabilities as well as voltage regulation, frequency control and other power quality issues [47,48]. UCs are currently available in many sizes. Two 3000 F capacitors are shown in Figure 6. There is even a startup company that claims to be able to create ultra capacitors with higher energy densities than lead-acid, Nickel metal hydride and even lithium ion batteries [49]. There have also been advances in the design of the ultracapacitors using nano-tube technology to improve the surface area of the capacitor. This "nano-tube ultracapacitor" would improve the ultracapacitor's energy density to be compatible with again that of a chemical battery [46,50,51].

4.3. Flywheel

During the past decade, flywheel energy storage systems (FESSs) have been rediscovered by the power industry due to their advantages in comparison with other energy storage systems. FESSs have found an important technical role on the application of enhancing the electric power quality, grid voltage and frequency support, and unbalanced load compensation. By virtue of their high dynamics, long lifetime, and good efficiency, FESSs are well suited for short-term storage systems [52-54].

FESS consists of a flywheel coupled to permanent magnets synchronous machine (PMSM) as shown in Figure 7. Flywheels store energy in the form of momentum in a rotating wheel or cylinder. A FESS stores energy through accelerating a rotor up to a high rate of speed and maintaining the energy in the system as inertial energy. Advanced composite materials are sometimes used for the rotor to lower its weight while allowing for the extremely high speeds. The flywheel releases the energy by reversing the process and using the motor as a generator. As the flywheel releases its stored energy, the

Figure 6. 3000 farad ultracapacitors.

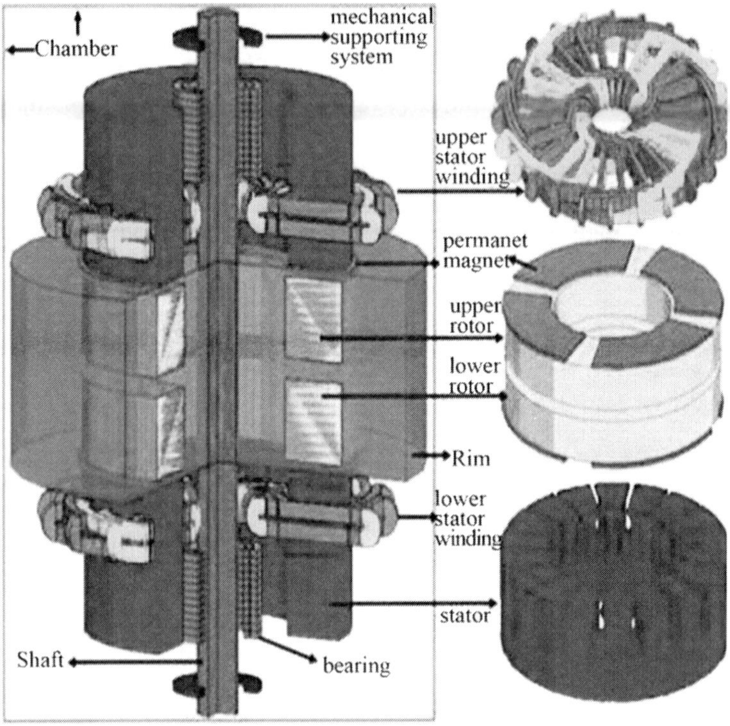

Figure 7. Flywheel energy storage system.

flywheel's rotor slows until it is fully discharged. The energy that can be stored depends on its rotational velocity ω and moment of inertia I, as in (3). They have high power density and more energy can be stored if the flywheel rotates at higher rotational speed [55]. Power electronics are used to ensure that output voltage has appropriate amplitude and frequency characteristics.

$$E = \frac{1}{2} I \omega^2 \tag{3}$$

Flywheel energy storage systems using superconducting magnetic bearing (SMB) together with a permanent magnet bearing (PMB) is one of the most promising electro-mechanical energy storage systems consisting long life, high energy density, high efficiency, with no pollution or toxic material disposal problems, and low rotational loss by non-contact superconductor bearing [55- 57].

Flywheels have seen most commercial success targeted for power delivery capabilities typically in the 150 kW-1MW range [50,58]. Recently, a 20 MW FESS plant has been successfully established in Stephentown, New York by Beacon Power Corp. under a pilot project of the Department of Energy (DOE), New York State Energy Research and Development Authority (NYSERDA) and currently 40 MW FESS plant project is under development [59].

4.4. Superconducting Magnetic Energy Storage (SMES)

Superconducting magnetic energy storage (SMES) systems store energy in the magnetic field produced by current flowing through a superconducting coil. The SMES principle is based on inductive energy storage in the magnetic field produced by current flowing through a superconducting coil.

A SMES system consists of four major subsystems [60], plus miscellaneous equipment for system control, data collection, and so on. The major subsystems are: 1) superconducting coil with magnet (SCM); 2) power conditioning system (PCS) that controls the flow of current into and out of the coil to charge and discharge the SMES; 3) cryogenic system (CS) that maintains the coil at a low enough temperature to maintain superconductivity; and 4) the control unit (CU), as shown in Figure 8.

For a SMES system, the inductively stored energy (E) and the rated power (P) are commonly the given specifications for SMES devices, and can be expressed as follows:

$$E = \frac{1}{2}LI^2 \; ; \; P = \frac{dE}{dt} = LI\frac{dI}{dt} = VI \tag{4}$$

where L is the inductance of the coil, I is the dc current flowing through the coil, and V is the voltage across the coil. During SMES operation, the magnet coils have to remain in the superconducting status. A refrigerator in the cryogenic system maintains the required temperature for proper superconducting operation [61].

Among the different variants of flexible alternating current transmission system (FACTS) devices and ESSs currently available, static synchronous compensators (STATCOM) integrated with SMES has been proposed as the most adequate for participating of the primary frequency control because of SMES has high efficiency (95% to 98% [61]) and rapid response to power demand [62-64]. Depending on the control loop of its power conversion unit and switching characteristics, the SMES system can respond very rapidly (MWs/milliseconds). The ability of injecting/absorbing real or reactive power can increase the effectiveness of the control, and enhance system reliability and availability. Comparing with other storage technologies, the SMES technology has a unique advantage in two types of applications: Power system transmission control and stabilization, and power quality improvement [65].

Several SMESs in the range of kWh to MWh scale have been already implemented for compensation of load/ generation fluctuation as well as energy storage [66]. Recently, 10 MVA/20 MJ SMES prototype has been tested at an actual power system including hydro power generators in order to compensate the fluctuating power load for a metal rolling factory [67].

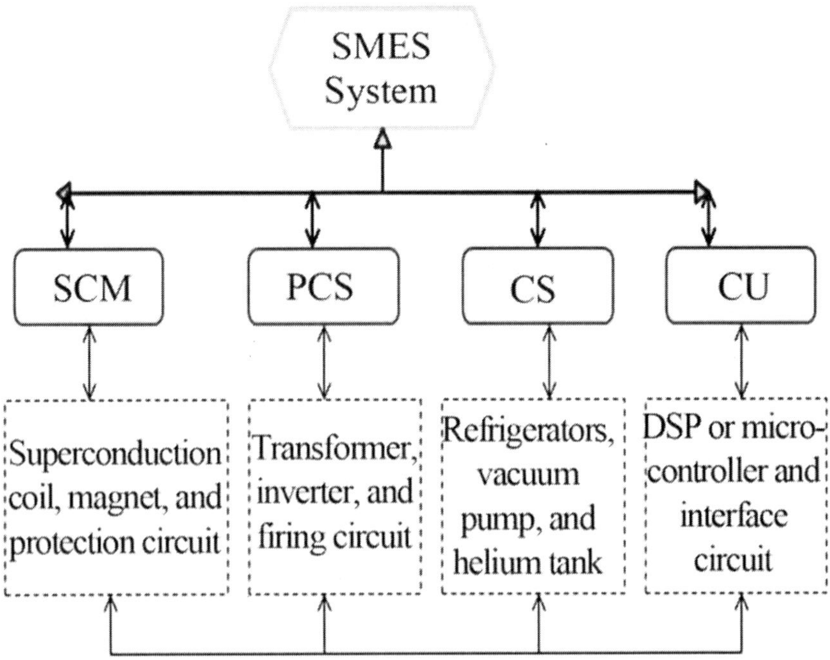

Figure 8. Block diagram of an SMES system.

4.5. Compressed Air Energy Storage System (CAES)

Compressed air energy storage system (CAES) is a hybrid technology of long term power storage and generation. A CAES plant mainly consists of 1) compressor train; 2) motor-generator unit; 3) gas turbine; and 4) underground compressed air storage tank (see Figure 9). During low-cost off-peak load periods, air is compressed and stored in large underground salt caverns [68]. Upon demand, the process is reversed; the compressed air is returned to the surface; this air is used to burn natural gas in the combustion chambers. The resulting combustion gas is then expanded in the two-stage gas turbine to spin the generator and produce electricity.

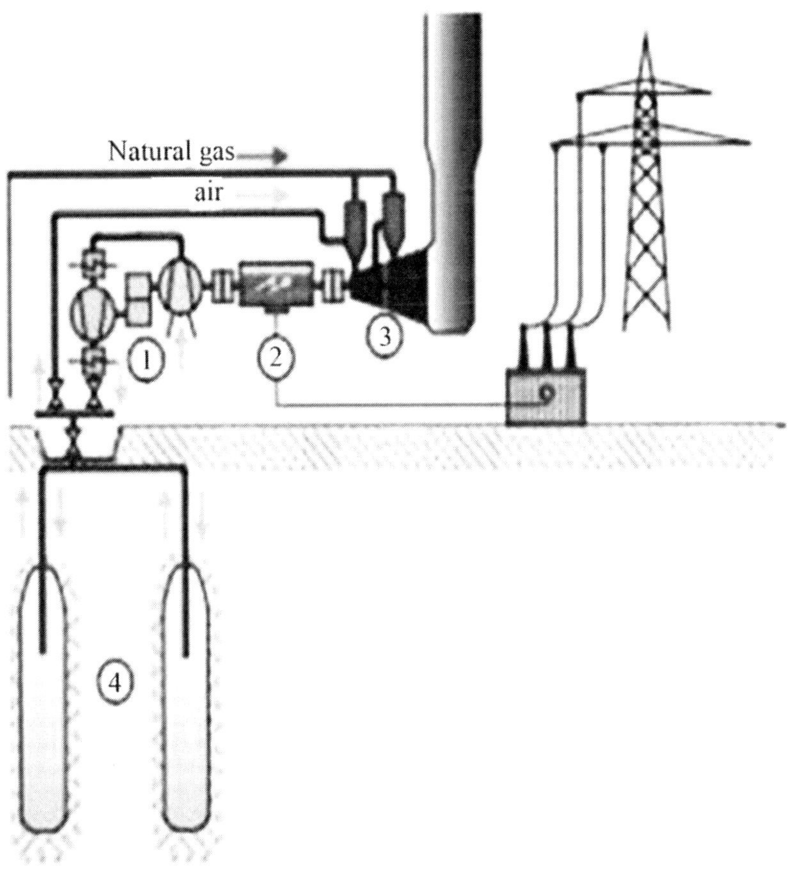

Figure 9. Major components of a CAES plant.

CAES can be placed near consumers and in large, small, and micro scale smart power networks. As in many countries, the potentials of installing new hydrostorage plants with pumping facilities are limited and many other energy storage devices are far from being economic, CAES may be an attractive investment opportunity for such purposes. The first such plant was built in Huntorf, Germany, in 1978 with a capacity of 290 MW [68]. In 1991, the first U.S. CAES facility was built in McIntosh, Alabama, by the Alabama Electric Cooperative and EPRI, and has a capacity rating of 110 MW [69]. Liu et al. [70] has proposed a novel hybridfuel CAES system for China. The design is based on using standard, industry proven equipment components to deliver a reliable and economic compressed air energy cycle. The resulting CAES plant consists of 410 MW generation, 205 MW of compression, and 2050 MW of storage.

In the past few years, research has been conducted to improve the efficiency of the turbines and heat transfer mechanisms used to pump and retrieve compressed air [71-73]. In an adiabatic CAES, the air's heat energy is stored separately and recovered before the compressed air is expanded in an air turbine. Such plants are currently under development and promise higher efficiencies and zero direct CO_2-emissions [71].

4.6. Pumped Hydroelectric Storage

Pumped hydroelectric storage (PHS) is the oldest kind of large-scale energy storage technology [74]. Since 1904, they are in active operation and new ones are still being built because of their operational flexibility and ability to provide rapid response to changes in system loading or spot price of electricity.

Conventional pumped hydroelectric storage consists of two large reservoirs, one is located at base level and the other is situated at a different elevation. Water is pumped to the upper reservoir where it can be stored as potential energy. Upon demand, water is released back into the lower reservoir, passing through hydraulic turbines which generate electrical power as high as 1000 MW. Pumped hydroelectric storage has huge energy and power capacity. Recently [75-77] examined the impact of pumped storage units together with large renewable penetration. They focused on reducing system operating costs and maximizing usage of renewable energy based on unit commitments and dispatches. Currently, efforts aimed at increasing the use of pumped hydro storage are focused on the development of underground facilities. Modern pumped storage plants are often designed to have fast start, load ramping and unloading capabilities. They can respond to load changes within seconds.

4.7. Hybrid Energy Storage System

The selection of the energy storage system for a particular application in power grid sometimes depends on a suitable combination of the power and energy ratings, energy, power density, cost, weight, volume, and operating temperature etc of ESSs. These requirements may not be achievable from a single energy storage source. To implement such optimal applications, hybrid energy storage devices (HESDs) have been proposed. In a HESD, two or more different energy storage systems with complementary characteristics can be combined together electronically. For future grid application proposed HESDs are listed next, with the energy-supplying device listed first, followed by the power-supplying device:

1) battery and ultracapacitor [78-81];
2) battery and flywheels [82];
3) CAES and battery or ultracapacitor [83];
4) battery and SMES [84].

Figure 10 demonstrates the relationship between power operational range and discharge time at rated power for various energy storage systems such as battery, ultracapacitors, flywheel, SMES, CAES, and pumped hydro. Table 2 summarizes the characteristic parameters of different energy storage technologies [30,42-44,59,83].

Table 2. Comparison of various energy storage systems.

Type	Energy Density (Wh/kg)	Energy Efficiency (%)	Power Density (W/Kg)	Life Time (Cycles)	Self Discharge (%/Month)	Environmental Effect
Battery	25 - 250	60 - 90	100 - 3000	150 - 3000	3 - 30	Toxic
Ultra Capacitor	<50	95	4000	10 k - 100 k	High	Benign
Flywheel	100 - 130	95	1000	>125,000	High	Benign
SMES	30 - 80	95	Very High	-	Negligible	Benign
CAES	10 - 30	50	Fair	40 Years	Negligible	Benign
PHS	0.3	65 - 80	-	75 Years	Negligible	Benign

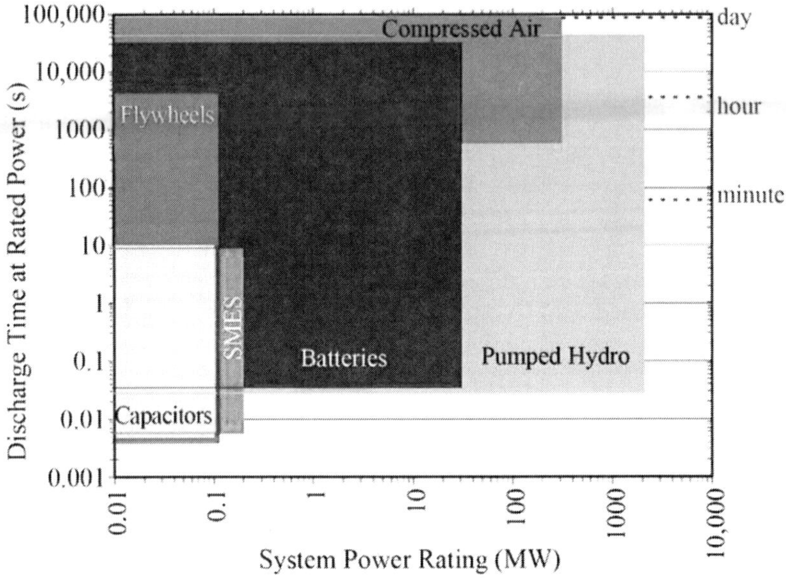

Figure 10. The relationship between power rating and discharge time for different storage technologies.

5. CONCLUDING REMARKS

In this paper the potential application of different ESSs and their prospect have been discussed and analyzed in detail for future smart power grids. Energy can be stored both for short and long durations. Various storage systems are available for these purposes. The importance of storage systems in electricity grids is finally receiving the attention of system planners as more storage options participation of storage is increasing. The design of smart grids in the future will take advantage of storage to deal with more dynamic loads and sources. As market rules are adjusted to take advantage of the benefits of bulk and distributed storage devices, the overall capabilities and reliability of more complex electricity networks should continue to improve as fully integrated smart grids.

REFERENCES

1. K. Moslehi and R. Kumar, "A Reliability Perspective of the Smart Grid," IEEE Transactions on Smart Grid, Vol. 1, No. 1, 2010, pp. 57-64.
2. P. Zhang, F. Li and N. Bhatt, "Next-Generation Monitoring, Analysis, and Control for the Future Smart Control Center," IEEE Transactions on Smart Grid, Vol. 1, No. 2, 2010, pp. 186-192.
3. S. M. Amin and B. F. Wollenberg, "Toward a Smart Grid: Power Delivery for the 21st Century," IEEE Power Energy Magazine, Vol. 3, No. 5, 2005, pp. 34-41.
4. International Energy Agency, "Technology Roadmap: Smart Grids," 2011. http://www.iea.org/publications/freepublications/publication /name,3972,en.html
5. US Department of Energy, "How the Smart Grid Promotes a Greener Future," 2010. http://energy.gov/oe/downloads/how-smart-grid-promotes-greener-future
6. H. Gharavi and R. Ghafurian, "Smart Grid: The Electric Energy System of the Future," IEEE Proceedings, Vol. 99, No. 6, 2011, pp. 917-921.
7. J. Xia and Y. Wang, "Secure Key Distribution for the Smart Grid," IEEE Transactions on Smart Grid, Vol. 3, No. 3, 2012, pp. 1437-1443.
8. A. Molderink, V. Bakker, M. G. C. Bosman, J. L. Hurink and G. J. M. Smit, "Management and Control of Domestic Smart Grid Technology," IEEE Transactions on Smart Grid, Vol. 1, No. 2, 2010, pp. 109-119.
9. A. Bose, "Smart Transmission Grid Applications and Their Supporting Infrastructure," IEEE Transactions on Smart Grid, Vol. 1, No. 1, 2010, pp. 11-19.
10. B. P. Roberts and C. Sandberg, "The Role of Energy Storage in Development of Smart Grids," IEEE Proceedings, Vol. 99, No. 6, 2011, pp. 1139-1144.
11. M. S. Whittingham, "History, Evolution, and Future Status of Energy Storage," IEEE Proceedings, Vol. 100, No. , 2012, pp. 1518-1534.
12. G. D. Rodriguez, "A Utility Perspective of the Role of Energy Storage in the Smart Grid," 2010 IEEE Power and Energy Society General Meeting, Minneapolis, 25-29 July 2010, pp. 1-2.

13. US Department of Energy, "Energy Storage—A Key Enabler of the Smart Grid," 2009. http://www.netl.doe.gov/smartgrid/refshelf.html #White%20Papers

14. US Department of Energy, "The Smart Grid: An Introduction, 2008. http://energy.gov/oe/downloads/smart-grid-introduction-0

15. F. Li, W. Qiao, H. Sun, H. Wan, J. Wang, et al., "Smart Transmission Grid: Vision and Framework," IEEE Transactions on Smart Grid, Vol. 1, No. 2, 2010, pp. 168-177.

16. M. D. Hopkins, A. Pahwa and T. Easton, "Intelligent Dispatch for Distributed Renewable Resources," IEEE Transactions on Smart Grid, Vol. 3, No. 2, 2012, pp. 1047- 1054.

17. S. S. S. R. Depuru, L. Wang, V. Devabhaktuni and N. Gudi, "Smart Meters for Power Grid—Challenges, Issues, Advantages and Status," 2011 IEEE/PES Power Systems Conference and Exposition (PSCE), Phoenix, 20-23 March 2011, pp. 1-7.

18. D. Rua, D. Issicaba, F. J. Soares, P. M. R. Almeida, R. J. Rei and J. A. P. Lopes, "Advanced Metering Infrastructure Functionalities for Electric Mobility," 2010 IEEE PES Innovative Smart Grid Technologies Conference Europe (ISGT Europe), Porto, 11-13 October 2010, pp. 1-7.

19. H. Suleiman, K. A. Ahmed, N. Zafar, et al., "Inter-Domain Analysis of Smart Grid Domain Dependencies Using Domain-Link Matrices," IEEE Transactions on Smart Grid, Vol. 3, No. 2, 2012, pp. 692-709.

20. Microsoft Power and Utilities, "Smart Energy Reference Architecture (SERA)," 2009.

21. A. P. S. Meliopoulos, G. Cokkinides, R. Huang, E. Farantatos, S. Choi, et al., "Smart Grid Technologies for Autonomous Operation and Control," IEEE Transactions on Smart Grid, Vol. 2, No. 1, 2011, pp. 1-10.

22. A. Ipakchi and F. Albuyeh, "Grid of the Future," IEEE Power Energy Magazine, Vol. 7, No. 2, 2009, pp. 52-62.

23. R. A. F. Currie, G. W. Ault, C. E. T. Foote, N. M. McNeill, et al., "Smarter Ways to Provide Grid Connections for Renewable Generators," IEEE PES General Meeting, Minneapolis, 25-29 July 2010, pp. 1-6.

24. Q. Pang, H. Gao and M. Xiang, "Design of Intelligent Terminal Unit for Smart Distribution Grid," Proceedings of the IEEE China International Conference on Electricity Distribution (CICED), Nanjing, 13-16 September 2010, pp. 1-6.

25. F. Rahimi and A. Ipakchi, "Demand Response as a Market Resource under the Smart Grid Paradigm," IEEE Transactions on Smart Grid, Vol. 1, No. 1, 2010, pp. 82-88.

26. E. Santacana, G. Rackliffe, L. Tang and X. Feng, "Getting Smart," IEEE Power Energy Magazine, Vol. 8, No. 2, 2010, pp. 41-48.

27. K. C. Divya and J. Østergaard, "Battery Energy Storage Technology for Power Systems: An Overview," Electric Power System Research, Vol. 79, No. 4, 2009, pp. 511- 520.

28. M. M. Chowdhury, M. E. Haque, M. Aktarujjaman, M. Negnevitsky and A. Gargoom, "Grid Integration Impacts and Energy Storage Systems for Wind Energy Applications—A Review," Proceedings of the IEEE PES General Meeting, San Diego, 24-29 July 2011, pp. 1-8.

29. S. C. Smith, P. K. Sen and B. Kroposki, "Advancement of Energy Storage Devices and Applications in Electrical Power System," IEEE PES General Meeting, Pittsburgh, 20-24 July 2008, pp. 1-8.

30. S. Vazquez, S. M. Lukic, E. Galvan, L. G. Franquelo, et al., "Energy Storage Systems for Transport and Grid Applications," IEEE Transactions on Industrial Electronics, Vol. 57, No. 12, 2010, pp. 3881-3895.

31. A. A. Thatte and L. Xie, "Towards a Unified Operational Value Index of Energy Storage in Smart Grid Environment," IEEE Transactions on Smart Grid, Vol. 3, No. 3, 2012, pp. 1418-1426.

32. W. Kempton and J. Tomic, "Vehicle-to-Grid Power Fundamentals: Calculating Capacity and Net Revenue," Journal of Power Sources, Vol. 144, No. 1, 2005, pp. 268-279.

33. A. Bracale, P. Caramia and D. Proto, "Optimal Operation of Smart Grids Including Distributed Generation Units and Plug in Vehicles," International Conference on Renewable Energies Power Quality (ICREPQ'11), Las Palmas de Gran Canaria, 13-15 April 2011, pp. 553-558.

34. M. M. Biswas, K. K. Das, I. A. Baqee, M. A. H. Sadi and H. M. S. Forhad, "Prospects of Renewable Energy and Energy Storage Systems in Bangladesh and Developing Economics," Global Journal of Researches in Engineering (GJRE), Vol. 11, No. 5, 2011, pp. 23-31.

35. Y. M. Cheng, Y. C. Liu, S. C. Hung and C. S. Cheng, "Multi-Input Inverter for Grid-Connected Hybrid PV/ Wind Power System," IEEE Transactions on Power Electronics, Vol. 22, No. 3, 2007, pp. 1070-1076.

36. J. M. Carrasco, L. G. Franquelo, J. T. Bialasiewicz, E. Galvan, R. C. P. Guisado, et al., "Power Electronic Systems for the Grid Integration of Renewable Energy Sources: A Survey," IEEE Transactions on Power Electronics, Vol. 53, No 4, 2006, pp. 1002-1016.

37. R. B. Schainker, "Executive Overview: Energy Storage Options for a Sustainable Energy Future," IEEE PES General Meeting, Palo Alto, June 2004, Vol. 2, pp. 2309- 2314.

38. C. A. Hill, M. C. Such, D. Chen, J. Gonzalez, et al., "Battery Energy Storage for Enabling Integration of Distributed Solar Power Generation," IEEE Transactions on Smart Grid, Vol. 3, No. 2, 2012, pp. 850-857.

39. J. McDowall, "Nickel-Cadmium Batteries for Energy Storage Applications," The 14th Annual Battery Conference on Applications and Advances, Long Beach, 12-15 January 1999, pp. 303-308.

40. J. P. Barton and D. G. Infield, "Energy Storage and Its Use with Intermittent Renewable Energy," IEEE Transactions on Energy Conversion, Vol. 19, No. 2, 2004, pp. 441-448.

41. P. C. Symons, "Opportunities for Energy Storage in Stressed Electrical Supply Systems," IEEE PES Summer Meeting, Vancouver, July 2001, Vol. 1, pp. 448-449.

42. D. Y. Goswami and F. Kreith, "Energy Conversion," CRC Press, Boca Raton, 2007.

43. A. A. H. Hussein, S. Harb, N. Kutkut, J. Shen and I. Batarseh, "Design Considerations for Distributed MicroSto- rage Systems in Residential Applications," 32nd International Telecommunications Energy Conference (INTELEC), Orlando, 6-10 June 2010, pp. 1-6.

44. B. R. Williams and T. Hennessy, "Energy Oasis [Vanadium Redox Battery System in Power Distribution Application]," IEEE Power Engineer, Vol. 19, No. 1, 2005, pp. 28-31.

45. Nesscap Ultracapacitors Products, "(EDLC) Electric Double-Layer Capacitor," 2012. http://www.nesscap.com/product/edlc_large1.jsp

46. D. Halber, "Researchers Fired Up over New Battery," MIT Tech Talk, Vol. 50, No. 16, 2006, pp. 1-5. http://web.mit.edu/newsoffice/2006/batteries-0208.html

47. Maxwell Technologies, "An Overview of Ultracapacitor Technology," 2012. http://www.maxwell.com/ultracapacitors/

48. R. D. Weir and C. W. Nelson, "Utilization of Poly (Ethylene Terephthalate) Plastic and Composition-Modified Barium Titanate

Powders in a Matrix that Allows Polarization and the Use of Integrated-Circuit Technologies for the Production of Lightweight Ultrahigh Electrical Energy Storage Units (EESU)," US Patent No. 7466536, 2008.

49. H. Pan, J. Li and Y. P. Feng, "Carbon Nanotubes for Supercapacitor," Nanoscale Research Letters, Vol. 5, No. 3, 2010, pp. 654-668.

50. H. Zhang, G. P. Cao, Y. S. Yang and Z. N. Gu, "Comparison between Electrochemical Properties of Aligned Carbon Nanotube Array and Entangled Carbon Nanotube Electrodes," Journal of Electrochemical Society, Vol. 155, No. 2, 2008, pp. 19-22.

51. R. Hebner, J. Beno and A. Walls, "Flywheel Batteries Come Around Again," IEEE Spectrum, Vol. 39, No. 4, 2002, pp. 46-51.

52. G. Cimuca, C. Saudemont, B. Robyns and M. M. Radulescu, "Control and Performance Evaluation of a Flywheel Energy Storage System Associated to a VariableSpeed Wind Generator," IEEE Transactions on Industrial Electronics, Vol. 53, No. 4, 2006, pp. 1074-1085.

53. G. Cimuca, S. Breban, M. Radulescu, C. Saudemont, et al., "Design and Control Strategies of an Induction-Machine-Based Flywheel Energy Storage System Associated to a Variable-Speed Wind Generator," IEEE Transactions on Energy Conversion, Vol. 25, No. 2, 2010, pp. 526-534.

54. M. Subkhan and M. Komori, "New Concept for Flywheel Energy Storage System Using SMB and PMB," IEEE Transactions on Applied Superconductivity, Vol. 21, No. 3, 2011, pp. 1485-1488.

55. J. Lee, S. Jeong, Y. H. Han and B. J. Park, "Concept of Cold Energy Storage for Superconducting Flywheel Energy Storage System," IEEE Transactions on Applied Superconductivity, Vol. 21, No. 3, 2011, pp. 2221-2224.

56. T. Ichihara, K. Matsunaga, M. Kita, I. Hirabayashi, et al., "Application of Superconducting Magnetic Bearings to a 10 kWh-Class Flywheel Energy Storage System," IEEE Transactions on Applied Superconductivity, Vol. 15, No. 2, 2005, pp. 2245-2248.

57. N. Hamsic, A. Schmelter, A. Mohd, et al., "Stabilising the Grid Voltage and Frequency in Isolated Power Systems Using a Flywheel Energy Storage System," The Great Wall World Renewable Energy Forum, Beijing, 23-27 October 2006, pp. 1-6.

58. E. Ortjohann, N. Hamsic, A. Schmelter, et al., "Increasing Renewable Energy Penetration in Isolated Grids Using a Flywheel Energy Storage System," International Conference on Power Engineering, Energy and Electrical Drives, Setubal, 12-14 April 2007, pp. 195-200.

59. M. Lazarewicz and J. Judson, "Performance of First 20 MW Commercial Flywheel Frequency Regulation Plant," ESA 2011 Annual Meeting, Beacon Power Corporation, San Jose, June 2011.

60. D. Sutanto and K. W. E. Cheng, "Superconducting Magnetic Energy Storage Systems for Power System Applications," International Conference on Applied Superconductivity and Electromagnetic Devices, Chengdu, 25-27 September 2009, pp. 377-380.

61. IEEE Task Force and T&D Committee, "Detailed Modeling of Superconducting Magnetic Energy Storage (SMES) System," IEEE Transactions on Power Delivery, Vol. 21, No. 2, 2006, pp. 699-710.

62. V. Karasik, K. Dixon, C. Weber, B. Batchelder, G. Campbell and P. Ribeiro, "SMES for Power Utility Applications: A Review of Technical and Cost Considerations," IEEE Transactions on Applied Superconductivity, Vol. 9, No. 2, 1999, pp. 541-546.

63. M. G. Molina and P. E. Mercado, "Comparative Evaluation of Performance of a STATCOM and SSSC both Integrated with SMES for Controlling the Power System Frequency," 2004 IEEE/PES Transmission and Distribution Conference and Exposition: Latin America, São Paulo, 8-11 November 2004, pp. 535-541.

64. M. G. Molina and P. E. Mercado, "New Energy Storage Devices for Applications on Frequency Control of the Power System Using FACTS Controllers," Proceedings of X ERLAC, Iguazú, 18-22 May 2003, pp. 1-6.

65. Y. Tatusuta, S. Koso, H. Abe, M. Urata, H. Ohsaki, et al., "Development of SMES for Power System Control," IEEE Transactions on Applied Superconductivity, Vol. 14, No. 2, 2004, pp. 693-698.

66. S. Nomura, T. Shintomi, S. Akita, T. Nitta, R. Shimada and S. Meguro, "Technical and Cost Evaluation on SMES for Electric Power Compensation," IEEE Transactions on Applied Superconductivity, Vol. 20, No. 3, 2010, pp. 1373- 1378.

67. Q. Li, "Status and Future Prospect of SMES for Grid Applications," Advanced Microgrid Concepts and Technologies Workshops, Brookhaven National Laboratory, Washington DC, 2012.

68. F. Crotogino, K. U. Mohmeyer and R. Scharf, "Huntorf CAES: More than 20 Years of Successful Operation," The Solution Mining Research Institute Spring Meeting, Orlando, 23-25 April 2001.

69. Electric Power Research Institute, "Compressed Air Energy Storage: 1994," EPRI Brochure BR-102936.

70. W. Liu, Y. Yang, W. Zhang, G. Xu and Y. Wu, "A Novel Hybrid-Fuel Compressed Air Energy Storage System for China's Situation," Proceedings of the International Conference of Efficiency, Cost, Optimization, Simulation Environmental Impact Energy Systems, Perugia, 26-29 June 2012, pp. 1-16.

71. C. Bullough, C. Gatzen, C. Jakiel, M. Koller, et al., "Advanced Adiabatic Compressed Air Energy Storage for the Integration of Wind Energy," Proceedings of European Wind Energy Conference, EWEC, London, 22-25 November 2004, pp. 1-8.

72. C. Linnemann and M. W. Coney, "The Isoengine: Realization of a High-Efficiency Power Cycle Based on Isothermal Compression," International Journal of Energy Technology and Policy, Vol. 3, No. 1-2, 2005, pp. 66-84.

73. Ridge Energy Storage & Grid Services L. P., "The Economic Impact of CAES on Wind in TX, OK and NM," Texas State Energy Conservation Office, Austin, 2005.

74. P. Breeze, "Power Generation Technologies," Elsevier, Boston, 2005.

75. P. Brown, J. Lopes and M. Matos, "Optimization of Pumped Storage Capacity in an Isolated Power System with Large Renewable Penetration," IEEE Transactions on Power Systems, Vol. 23, No. 2, 2008, pp. 523-531.

76. J. G. Gonzalez, R. D. L. Muela, L. Santos and A. Gonzalez, "Stochastic Joint Optimization of Wind Generation and Pumped-Storage Units in an Electricity Market," IEEE Transactions on Power Systems, Vol. 23, No. 2, 2008, pp. 460-468.

77. A. Tuohy and M. O'Malley, "Impact of Pumped Storage on Power Systems with Increasing Wind Penetration," IEEE Power & Energy Society General Meeting, Calgary, 26-30 July 2009, pp. 1-8.

78. W. Henson, "Optimal Battery/Ultracapacitor Storage Combination," Journal of Power Sources, Vol. 179, No. 1, 2008, pp. 417-423.

79. H. Yoo, S. K. Sul, Y. Park and J. Jeong, "System Integration and Power Flow Management for a Series Hybrid Electric Vehicle Using Super-Capacitors and Batteries," IEEE Transactions on Applied Superconductivity, Vol. 44, No. 1, 2008, pp. 108-114.

80. L. Shuai, K. A. Corzine and M. Ferdowsi, "A New Battery/Ultracapacitor Energy Storage System Design and Its Motor Drive Integration for Hybrid Electric Vehicles," IEEE Transactions on Vehicular Technology, Vol. 56, No. 4, 2007, pp. 1516-1523.

81. M. B. Camara, H. Gualous, F. Gustin and A. Berthon, "Design and New Control of DC/DC Converters to Share Energy between Supercapacitors and Batteries in Hybrid Vehicles," IEEE Transactions on Vehicular Technology, Vol. 57, No. 5, 2008, pp. 2721-2735.

82. O. Briat, J. M. Vinassa, W. Lajnef, S. Azzopardi and E. Woirgard, "Principle, Design and Experimental Validation of a Flywheel-Battery Hybrid Source for Heavy-Duty Electric Vehicles," IET Electric Power Applications, Vol. 1, No. 5, 2007, pp. 665-674.

83. S. Lemofouet and A. Rufer, "A Hybrid Energy Storage System Based on Compressed Air and Supercapacitors with Maximum Efficiency Point Tracking (MEPT)," IEEE Transactions on Industrial Electronics, Vol. 53, No. 4, 2006, pp. 1105-1115.

84. T. Ise, M. Kita and A. Taguchi, "A Hybrid Energy Storage with a SMES and Secondary Battery," IEEE Transactions on Applied Superconductivity, Vol. 15, No. 2, 2005, pp. 1915-1918.

8

CHAPTER

OPTIMIZATION OF THERMAL AND ELECTRICAL APPLIANCE LOADS IN RESIDENTIAL AND COMMERCIAL BUILDINGS WITH DEMAND SIDE MONITORING

Emmanuel O. B. Ogedengbe

Energhx Research Group, Department of Mechanical Engineering, Faculty of Engineering, University of Lagos, Lagos, Nigeria.

ABSTRACT

Energy demand-supply mechanism with the load shape for both residential and commercial buildings in the Province of Ontario and Nigeria is studied with demand side monitoring of energy consumption. Thermal and electrical loads are characterized by certain predictor variables, including the consumers' behavioural pattern, power ratings of energy appliances and weather conditions. The proposed bottom-up approach is capable of providing low-volume electricity and natural gas consumers, in a fully deregulated energy market, with competitive energy saving advantage, based on corrective monitoring of independent users' demand loads. Special application of the bottom-up model-based facility characterization of demands for thermal comfort and indoor air qualityin a developing energy sector like Nigeria enables the development of planning tool for the proposed integration of renewable power systems. The developed DSMonitor™ app is capable of deploying an effective smart grid technology tool towards an improved building energy demand-supply balance at the individual end-user level.

Keywords: Optimization; Polynomial Regression; Electricity; Natural Gas; Energy Stocks; DSMonitor™; Smart Grid

1. INTRODUCTION

Efficient control of the demand-supply mechanism has often enabled the effective monitoring of the baseload generation of power by all stakeholders in the energy sector. Surplus baseload generation of electricity affects the prices of electricity in Ontario, leading into negative prices during hours when Ontario sellers pay buyers in the US or Quebec to take the excess electricity [1]. Generally, electricity prices are determined by various demand-supply mechanisms, including conservation procurement, delivered comparison, generation/production procurement, global adjustment, and the Ontario Energy Board's Regulated Price Plan (RPP). But with the new Feed-In-Tariff (FIT) programs of the Ontario Power Authority (OPA), the generation procurement mechanism is playing a prominent role in the determination of electricity price. As more green energy from wind and solar systems are procured, low-volume consumers seem left out from the attractive FIT programs, being at the receiving end of the hike in the effective electricity rates.

Since Ontario wholesale generators are guaranteed fixed payments under certain long-term contracts, the task of controlling the electrical grid system against surplus power becomes enormous for the province's Independent Electrical System Operator (IESO). Top-bottom prediction of the demand-supply forecast, based on econometric models and weather scenarios, is utilized by the IESO to control the demand-supply balance. Although this approach is most effective and reliable in a fully deregulated energy market with appreciable control of the demand-supply balance, the increase in the cost of surplus power suggests the development and utilization of auxilliary technology at the buidling level. When energy demand is negative, an economic consideration implies that an effective model that is capable of limiting the generation of surplus electricity will be desirable. Lund et al. [2] argued that while the identification of the problem of mismatch between energy demand and production of zero energy building is feasible at the building level, compensation of this mismatch at the same level is complicated. However, recent advances in smart grid technologies [3,4], and the quest for effective compensation of this mismatch challenge, require the development of a robust demand side monitoring tool at the building level in addition to the aggregate treatment of distributed mismatches.

Apart from the IESO, baseload generators, being the significant Market Participants in the Ontario energy market, are also capable of managing the appropriate balance of supply when there is surplus baseload generation. In a fully deregulated energy sector like Ontario, where low-volume consumers of natural gas and electricity (for example, the Steamed Rice Express Restaurant in Figure 1) have choices of deciding their energy supplier, the proposed demand side monitoring of the performance of domestic and non-domestic energy stocks offers an economic saving advantage, based on smart monitoring of energy consumption. This methodology is useful for management of surplus baseload generation with the decentralized energy planning system. However, the decentralized energy planning system is popular only in a developing economy like Nigeria, where supply even in cities is always negative and erratic. Hiremath et al. [5] developed a Goal Programming (GP) model with a decentralized energy planning (DEP) of energy demands in Tumkur District, using a bottom-up approach. DEP method enables the optimization of multi-objective resource allocation problems. With the royal assent on the Green Energy Act in Ontario in 2009, energy supply from large-scale deployment of solar photovoltaic (PV) technology contributes significantly to the problem of surplus baseload generation. The pioneering works of Denholm and Margolis [6] examined the limits of large-scale deployment solar PV for any utility grid system. As reviewed by Solomon et al. [7], their work proposed various options to improved solar penetration, including energy storage, energy dumping, increasing load flexibility, and load shifting. Unfortunately, Ontario's surplus electricity, sold to buyers in the US and the Province of Quebec, had always resulted in negative prices. This penalty was up sharply from the same period in 2010, when there were only 10 hours of negative prices at a cost of $4.2 million to $35 million in 2011 during the 95 hours of negative prices [8]. Without prejudice to the incentives through the Feed-in-Tariff (FIT) and the Micro Feed-in-Tariff (microFIT) programs, clean energy practice starts with effective conservation of the available electricity in the local grid and current natural gas production. Stemming the drive towards the estimated increase in the demand of electricity and natural gas to residential and commercial facilities plays a significant role in their respective price determination mechanisms. A global effort towards demand side management, remodeled by Energhx® and referred to as demandside monitoring (DSM) in this paper, has been reported by the International Energy Agency [9]. Figure 2 shows three-mode energy contracting with Energhx® towards improvement in building energy efficiency, including Solar Energy Supply Contracting (Solar ESC), Energy Supply Contracting (ESC) and Energy Performance Contracting (EPC). Residential and commercial facilities across the Province of Ontario are adopted as the scope for the DSM study. Unlike the practice of most conventional developer of solar photovoltaic system, this paper demonstrates the importance of DSM in order to address the problem of grid flexibility and surplus baseload generation.

Figure 1. Pictorial view of steamed rice express restaurant, a typical commercial energy system located in the city of Ottawa, Ontario.

Second Law analysis [10], based on exergetic design of the residential and commercial energy systems, is adopted for the design of thermal comfort, indoor air quality (IAQ) and performance variables of renewable energy system within each unit of the energy model [11]. Considered as an open system, analytical tool from the Natural Resources Canada (NRCan), including HOT2000, is considered alongside the DSMonitor™. As described by Filippi [12] and reported in Ref [13], there are four methodologies for assessing the primary energy demand in buildings. The first method involves the use of computational fluid dynamics tools, like the DSMonitor™, BLAST, DOE-2, EnergyPlus, etc. Although this method is complex and time-consuming, realistic solutions to energy performance and indoor air quality problems would be indeterminate without this approach [14]. The second method is a simplified approach, where the energy demand is evaluated assuming the building operates in standard conditions. The third method involves the use of simplified evaluation of primary energy demand. The fourth method is only applicable to real buildings, involving the analysis of the historical consumption data; and corrective intervention in order to improve the system efficiency of the building. This paper proposes the integration of the first and the last methods, involving both energy supply contracting and energy performance contracting.

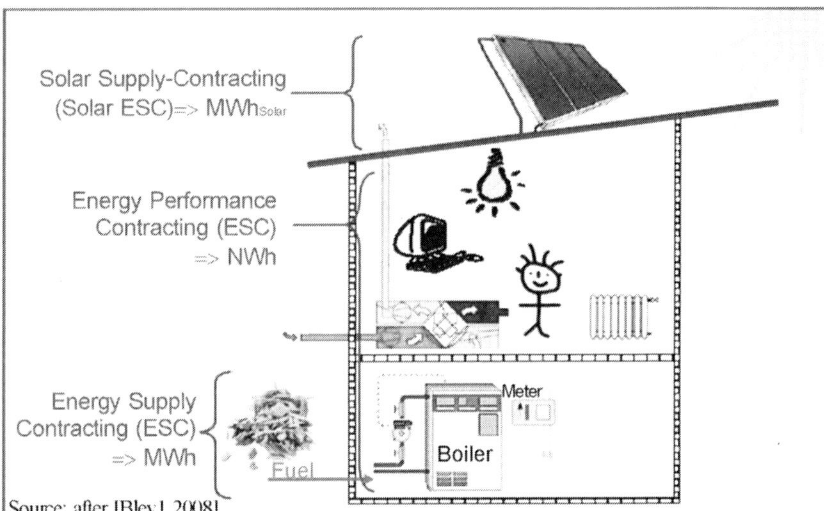

Figure 2. Schematic showing the two basic energy contracting models (Source: IEA).

2. DESCRIPTION OF AN ENERGY STOCK

In classical thermodynamics, every residential or commercial building facility for demand side monitoring and/or renewable energy project, like the engineering lecture theatre within the Faculty of Engineering, University of Lagos shown in Figure 3, are considered as an energy stock. The characterization of each of the energy stock based on thermal comfort, indoor air quality and the electrical appliances is grouped according to the type of facility (e.g., lecture theatres, offices, workshops/ laboratories, cafeteria, etc.).

The Second Law of Thermodynamics defines the concept of irreversibility, thereby necessitating the development of energy efficiency technology that can reduce the rate of exergy destruction in the energy stock [15]. Each of the components in this stock operates by drawing fuel, which is either natural gas or electricity commodities, in an open system. For example, a domestic hot-water system can operate with either natural gas as fuel or electricity, in order to generate the desired heat energy required to heat up the water for domestic use. With many other components within an energy stock, exchanges of internal and kinetic energies constitute the flow of irreversibility and reduction of system efficiency. Ogedengbe et al. [16] had covered similar modeling of energy exchanges at the wall of a micro-fluidic device, where exergy losses are contributed to slip-flow irreversibility at the near-wall of the micro-channel. This DSM study corroborates the same statement of the Second Law approach that energy conversion, between two different levels of state, introduces irreversibility; and only sensitivity analysis can reveal the optimal energy system efficiency. It is capable of leading to an improvement in the thermal comfort and indoor air quality, reduction in the consumption of electricity and natural gas, and recommendation of efficient energy generation system [17].

Figure 3. Roof-Top view of the faculty of engineering, university of Lagos.

An energy stock's control volume comprises the external boundary of a building envelope. Figure 4 shows a control volume based energy stock, modeled with threedimensional hexahedral finite elements. The boundary of this building envelope encloses a mass of fluid, representing transport of heat and fluid transport with significant energy conversion mechanisms. Assuming incompressible flows, these scalar transport variables can be predicted based on the following three-dimensional form of mass, momentum, indoor air concentration, and energy equations [18]

$$\frac{\partial(\rho)}{\partial t}+\frac{\partial(\rho U)}{\partial x}+\frac{\partial(\rho V)}{\partial y}+\frac{\partial(\rho W)}{\partial z}=0 \tag{1}$$

$$\frac{\partial(\rho U)}{\partial t}+\nabla\cdot(\rho v U)=-\frac{\partial P}{\partial x}+\nabla\cdot(\mu\nabla U)+\dot{S}_U \tag{2}$$

$$\frac{\partial(\rho V)}{\partial t}+\nabla\cdot(\rho v V)=-\frac{\partial P}{\partial y}+\nabla\cdot(\mu\nabla V)+\dot{S}_V \tag{3}$$

$$\frac{\partial(\rho W)}{\partial t}+\nabla\cdot(\rho v W)=-\frac{\partial P}{\partial z}+\nabla\cdot(\mu\nabla W)+\dot{S}_W \tag{4}$$

$$\frac{\partial(\rho c_p T)}{\partial t}+\nabla\cdot(\rho c_p v T)=\nabla\cdot(k\nabla T) \tag{5}$$

$$\frac{\partial(\rho c_p C)}{\partial t}+\nabla\cdot(\rho c_p v C)=\nabla\cdot(k\nabla C) \tag{6}$$

2.1. Demand-Supply Mechanism

The Ontario Electronic Business Transaction [19] ("EBT") standard for electricity retailing and the EBT Data Transport Protocol [20] for natural gas marketing recommend the Public Key Infrastructure technology ("PKI") as the key standard to adopt in ensuring EBT Internet security characteristics of Privacy, Authentication, Integrity, and Non-repudiation ("PAIN"). The Meter Data and Payments Transactions in Figure 5 represent the required information in order to capture the historical consumption of every ESCo Client. The inclusion of the historical usage data, for up to twenty four (24) months, represents the significant contribution of this paper to the monitoring of energy commodities supplied to the energy stock. Figure 6 shows the schematic of the energy system loads, including the energy building demand and the energy component supply. Without the analysis of the energy component supply, Sakulpipatsin et al. [21] had considered the application of exergy in the analysis of the thermal and electricity demands on the building and services that makes up the energy building system. A onezone model for the space energy balances was considered, assuming all walls are adjacent to the surrounding environment.

2.1.1. Energy Building Demand

Energy demand within every building facility is divided into two categories, including building demand and service demand. Figure 6 shows the schematic of the distributions and the component supplies. The building demand is utilized for thermal conditioning of the building envelope; while the service demand takes care of the indoor air conditioning, thermal comfort control, and the operation of all appliances. The building energy balance becomes

$$\frac{DQ_{\text{air}}}{dt} = Q_{\text{heating}} - Q_{\text{cooling}} + Q_{\text{infiltration}} + Q_{\text{ventilation}} + Q_{\text{transmission}} + Q_{\text{gain}} \quad (7)$$

$$\frac{DQ_{\text{surface}}}{dt} = Q_{\text{combined,o}} - Q_{\text{combined,i}} + Q_{\text{radiation,i}} + Q_{\text{radiation,o}} \quad (8)$$

The limitation of one-zone model and other approaches where simplified evaluation of the primary energy demand is adopted is that they are not reliable for indoor quality analysis; and inadequate in region where the ranges of weather changes throughout the year play a significant role in the performance of energy systems.

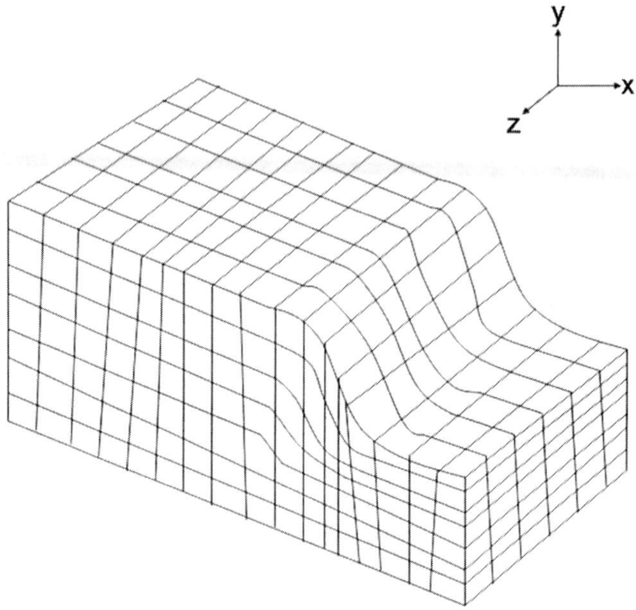

Figure 4. Control volume based energy stock, modeled with 3-D hexahedral finite elements.

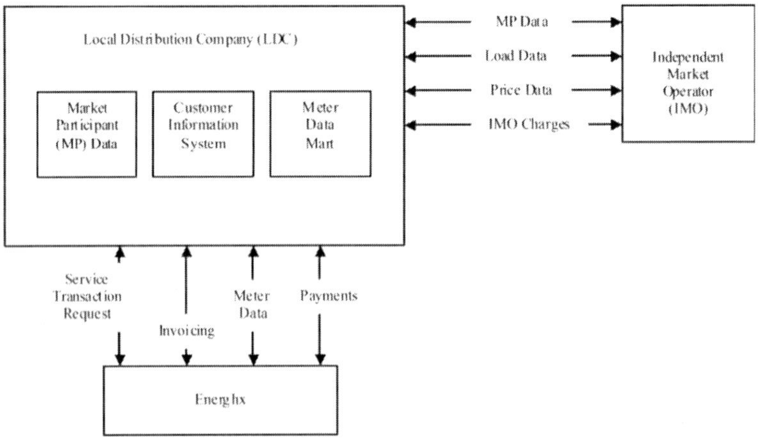

Figure 5. Ontario EBT standards for exchange of information between suppliers and LDC.

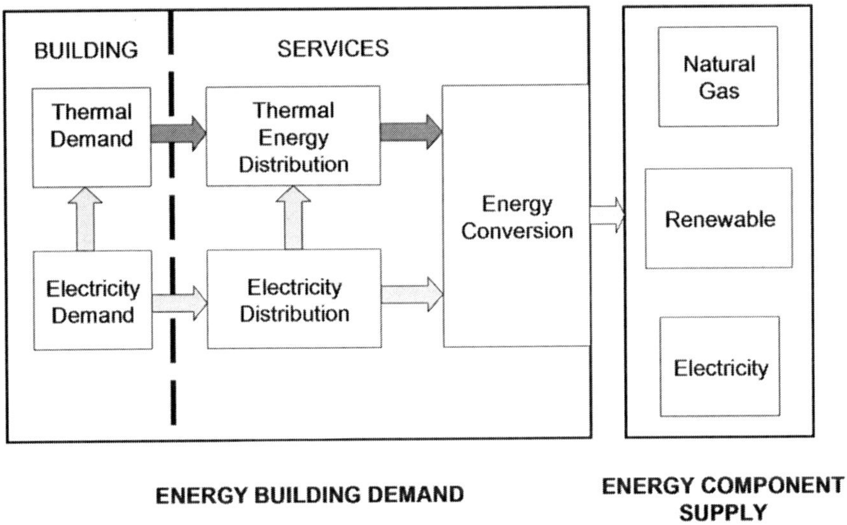

Figure 6. Schematic of the distribution of energy building demand and the energy component supply.

2.1.2. Energy Commodity Supply

The supply of electricity and natural gas in the Ontario energy market is fully deregulated. The deregulation of the electricity market, which removes the monopoly of the former Ontario Hydro, came into force in May, 2002; while the opening of the market for natural gas occured in 1986. Apart from the positive effect of deregulation on strengthening of the economic and the encouraging search for alternative sources of power, low-volume consumers have options of choosing their energy carrier from either the marketers or their local distribution companies (LDC). However, the service offers from most of the energy carriers have been limited to the marketing of price derivative supply of electricity and natural gas. Consumers merely have to choose between various marketers-determined fixed price protection plans for a maximum of five years term, or remain on the Ontario Energy Board (OEB)-determined regulated price plan, known as the standard supply service offered by their local distribution companies. While a price-derivative service offers protection against volatility of prices in the energy market, sustainablity of energy and the environment may strongly depend on alternative supply of energy based on consumption-derivative services.

2.2. Renewable Energy Supply and Integration

The concept of renewable energy supply as compensation factor at the building level represents a significant component of the bottom-up modelling approach. The compensation factor reflects the specific type of renewable energy conversion technology involved, among solar photovoltaic (solar PV), wind, biomass, etc. For example, solar PV cells/modules convert solar irradiation from the sun into direct current of electricity. With the proposeddemand side monitoring of energy stocks, the grouping of series of solar PV modules is designed to meet the expected system requirement and the available space for installation. Efficient harvest of the energy from sunlight is possible when the modules are oriented between southeast and south-west. For a typical distributed solar PV system designed for a rooftop installation in the City of Ottawa, Table 1 presents the required component descriptions for snow and wind loads calculation.

Conventional design consideration without the audit of the energy building demand leaves far too many uncertainties about the economic and sustainable viability of the proposed distributed generation project. Figure 6 shows a schematic of the energy demand and distribution within the system and proposed supply mix. The total design load calculated in Table 1 on the basis of roof zone, external wind and snow loads do not consider the demand loads of the consumer. Its capacity is beyond the consideration of independent, power generation and consumption.

3. DEMAND SIDE MONITORING

The commercial building ("C1-Ottawa") is located in the City of Ottawa, while the residential buildings, i.e., R1- Guelph and R1-Ottawa, are located in the City of Guelph and the City of Ottawa, respectively. Energy supplies to these buildings, including electricity and natural gas, are from Energhx®. Tables 2, 3 and 4 show the physical properties of these energy stocks.

Energy consumption values were simulated for each of the components within the three energy stocks [22]. The results of the simulation are summarized in Tables 5, 6, and 7. Due diligence was exercised to obtain input data for the HOT2000 simulation software, using one or more of: a nominal power rating in the appliance's manual; an Energuide energy consumption per year rating; and the use of a Kill-A-Watt device. The Kill-A-Watt electricity usage monitor (P3 International, Model P4400) is actually utilized to capture the consumption data for the R1-Guelph stock as shown in Table 6. By plugging any appliance into the Kill-A-Watt, the instantaneous readout of its power rating, as well as a running total of energy consumed, are displayed.

Table 1. Component considerations for design loads calculation.

Variable description	Symbol	Value	Unit
Building height	H	5.486	m
Building, least horizontal dimension	B	6.8	m
	Downforce (cases)		
Description (variable/unit)	**1**	**2**	**3**
Dead load (D/Pa)	134.71	134.71	134.71
Snow load (S/Pa)	2602.69		1952.01
Design wind load (P_{net}/Pa)		254.34	190.75
Total design load (P/Pa)	2737.40	389.04	2277.48

Table 2. Physical properties of C1-Ottawa building envelope elements.

Element	Qty	Orientation	Area [m^2]	U-value [$W/(m^2 \cdot K)$]
Wall	10	N/S/E/W	372.38	0.2
Floor	2		99.96	0.31
Window	4	West and East	6.89	0.025
Ceiling	2		99.8	0.31

Table 3. Physical properties of R1-Guelph building envelope elements.

Element	Qty	Orientation	Area [m²]	U-value [W/(m²·K)]	g-value
Exterior Walls	4	North	79.58	0.412	
		East	69.97	0.384	
		South	56.38	0.407	
		West	61.76	0.396	
Basement	1		71.11	1.82	
Windows	4	North	7.97	4.83	0.622
		East	1.54	4.92	0.562
		South	7.16	4.90	0.635
		West	0.41	5.59	0.597
Ceiling	1		95.79	0.348	

Table 4. Physical properties of R1-Ottawa building envelope elements.

Element	Qty	Orientation	Area [m²]
Main Walls	13	East	118.87
		west	108.22
		North	78.47
		South	86.35
Window	18	East	3.38
		west	6.15
		North	5.98
		South	14.84
Floors	3	Main	138.3
		Second	134.85
		Basement	135.72
Ceiling	2	Ceiling 1	161.5
		Ceiling 2	95.17
Doors	3	Door1	3.31
		Door 2	3.11
		Door 3	2.99

Table 5. Annual energy consumption for C1-Ottawa.

Element	Qty	Orientation	Area [m^2]	U-value [W/(m^2·K)]
Wall	10	N/S/E/W	372.38	0.2
Floor	2		99.96	0.31
Window	4	West and East	6.89	0.025
Ceiling	2		99.8	0.31

Table 6. Annual energy consumption for R1-Guelph (E = electricity; NG = natural gas).

Component	Fuel type	Consumption (kWh or m^3)	Component	Fuel type	Consumption (kWh or m^3)
Space heating	NG	1733	Washer	E	26
DHW heating	NG	384	Dryer	E	189
A/C unit	E	3285	Family TV	E	228
Water softner	E	30	L-level TV	E	5
Freezer chest	E	646	U-level TV	E	16
Fridge	E	464	Desktop	E	1235
Microwave	E	210	Fans	E	2102
Range	E	876	CODetector	E	61
Toaster oven	E	219	Lighting	E	126
Water boiler	E	129	Total	NG	2117
			Total	E	9848

Table 7. Annual energy consumption for R1-Ottawa (E = electricity; NG = natural gas).

Component	Fuel type	Consumption (kWh or m^3)
DHW heating	NG	2914.96
Space heating	NG	1599.07
Space cooling	E	16475.57
Appliances	E	13305.45

HOT2000 presents the energy consumption profiles, based on behavioural pattern, weather condition, and the distribution of energy appliances, not only at an instance but along a time distribution throughout the time of year. The models of energy consumption by device are shown in Figures 7, 8, and 9. Comparison of these distributions with historical consumption profiles can provide a useful analytical tool for the development of an independent energy demand model. The simulation predicted a consistent, near-parabolic profile for the energy demand for

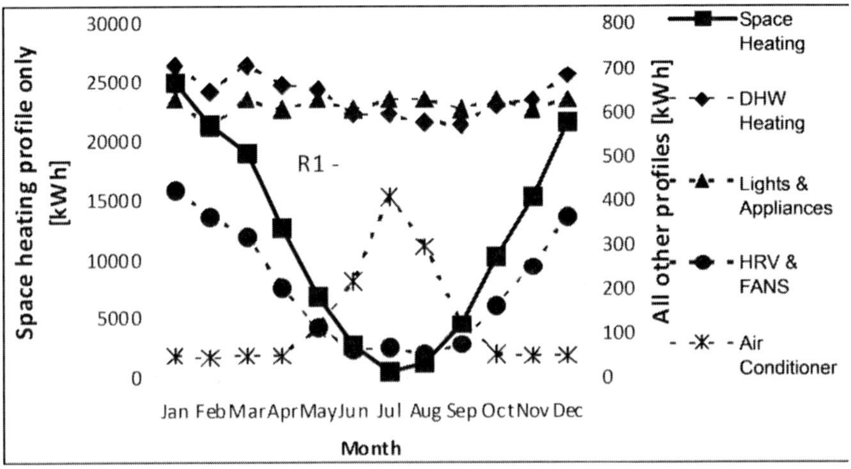

Figure 7. The monthly estimated energy consumption (by device) for the R1-Guelph stock.

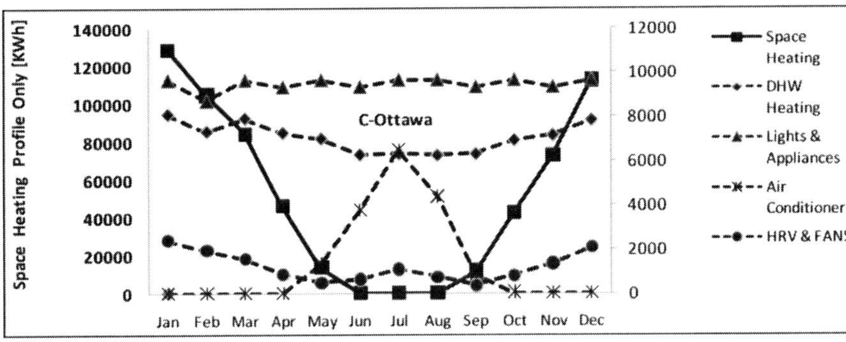

Figure 8. The monthly estimated energy consumption (by device) for the C1-Ottawa stock.

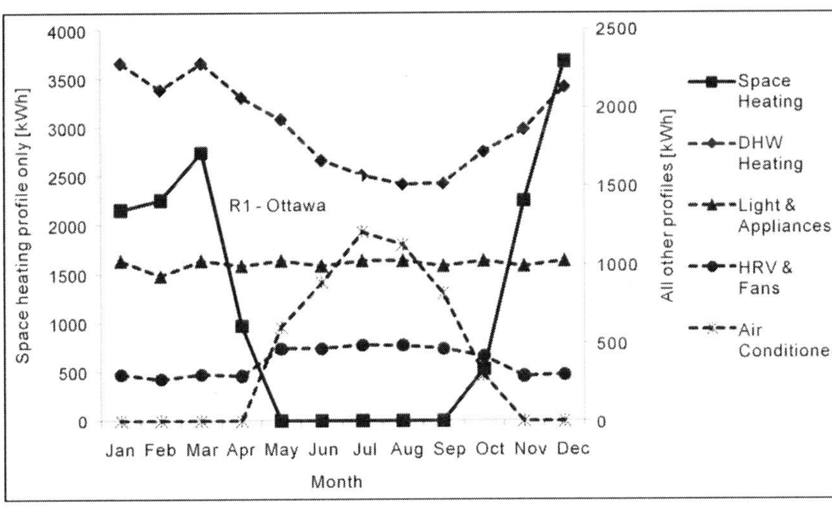

Figure 9. The monthly estimated energy consumption (by device) for the R1-Ottawa stock.

space heating for the three energy stocks. This shows a convergence in the behavioral response to the demand for space heating, and about the same weather condition prevailing in both the City of Ottawa and the City of Guelph during the period of this study. Since the fuel for space heating for these stocks is natural gas, difference in the consumption profile for natural gas would depend of the DHW Heating. Coincidentally, and unlike the DHW Heating demand for R1-Guelph, the DHW Heating demand profiles for both R1-Ottawa and C1-Ottawa appear identical.

Energy demand for space cooling peaks uniformly around the month of July for the three energy stocks considered in this study. However, a significant decrease in the gradient of the peak is noticed for the R1-Ottawa, showing more in the consumption of electricity for space cooling (see Figure 9 and Table 7). Since the proposed corrective measures toward the proposed energy saving recommendations of this study are not aimed at building retrofit services, default values in HOT2000 are used for the wall configuration, ceiling configuration, and floor configuration.

3.1. Facility Characterization

The historical consumption data for 24 months, providing a significant study of the repetition of the seasonal influence, is assessed from the local distribution companies. Figures 10, 11 and 12 show the historical consumption of natural gas and electricity for the three energy stocks. Significant peak in the consumption of natural gas duringthe winter period underscores the increase in demand for space heating and hot-water heating. Caldera et al. [9] studied the influence of the building age on the space heating demand of residential building in Torino. Both linear and power interpolations were used to obtain the statistical correlations of the building stocks. With visual inspection of the profiles over two consecutive seasons, this DSM study shows that the pattern of the consumption of natural gasis consistent for both R1-Ottawa and R1-Guelph. However, it appears that there is a significant behavioural change within this period of study that affects only the consumption of electricity. While a peak in demand for electricity both the summer and winter for R1-Guelph is attributed to space cooling and HRV & Fan respectively, both the domestic and non-domestic energy stocks in the City of Ottawa show different profiles. For example, occupancy pattern is a strong factor that influences the consumption of electricity for HRV & Fans, and lighting & appliance demands. Widen et al. [23] uses Markov chain model to generate the occupancy patterns and daylight availability in Swedish households.

Seeing that the distribution of the energy needed in a building includes heating, cooling, ventilation, hot water, lighting, and various electric appliances [24]; the complexity of the distribution demands independent modeling of individual energy stock.

3.2. Bottom-Up Modelling Approach

The major difference between cost-based model and consumption-based model is that the former depends on the conventional model for price forecasting of energy commodity, which is tied to the top-down or aggregate analysis of the market; while the consumption-based model depend on the characteristics of the end-users'facilities. Significant among the top-down models are the time series and simulation models, including the market assssment and portfolio strategies (MAPS) algorithm and the UPLAN software [25]. However, the proposed bottom-up modelling approach is strictly a

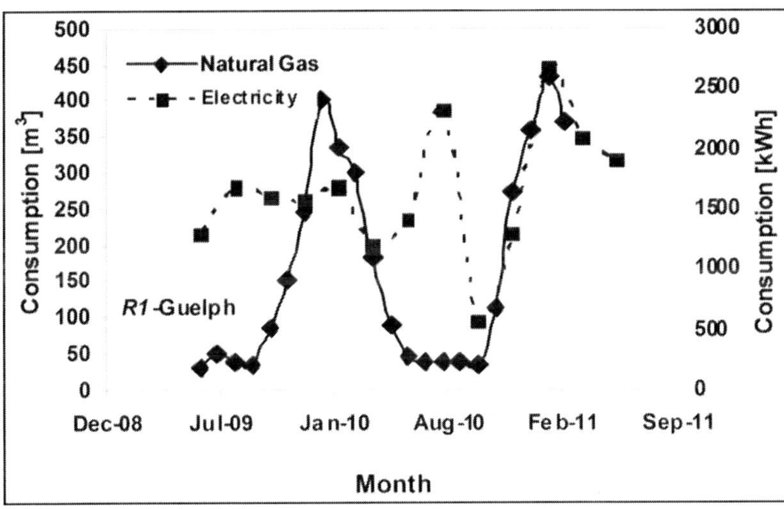

Figure 10. Electricity and natural gas load shape for R1-Guelph stock.

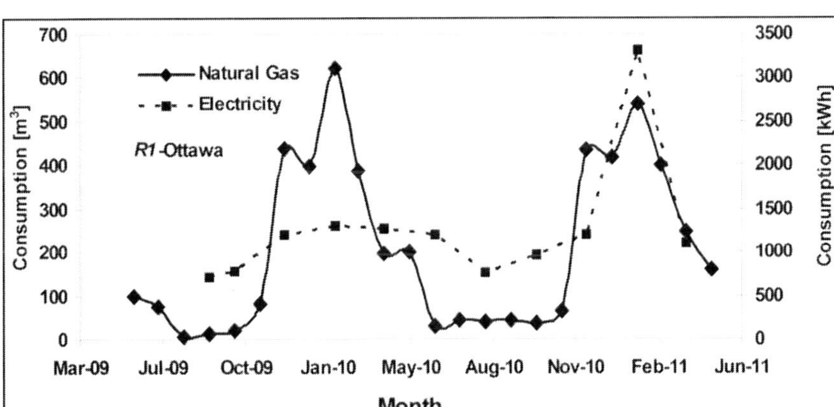

Figure 11. Electricity and natural gas load shape for R1-Ottawa stock.

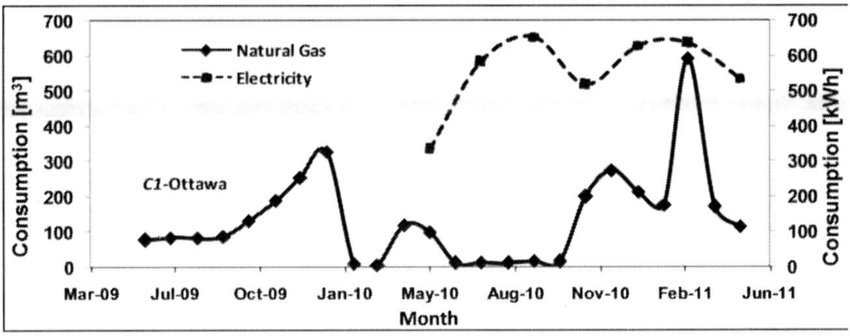

Figure 12. Electricity and natural gas load shape for C1-Ottawa stock.

time-series model, based on the characterization of the facility demands for thermal comfort, indoor air quality and electrical appliance loads. Aggrawal et al. [24] identified three types of time-series models, including the parsimonious stochastic model, artificial intelligence model and the regression model. While the stochastic model is characterized by univariate discrete analysis with autoregressive modelling of energy price spikes and mean reversion, the artificial intelligence models are non-parametric models that map the input-output relationships without exploring the underlying process. Although, the regression model is adopted in this study, the predictor variables are characterized with the thermal comfort, indoor air quality and electrical appliance load at the individual energy stock or building level.

Proposing a m-predictor-variable regression analysis, including the indoor concentration of CO_2, occupancy, weather, wattage and usage frequency of electrical appliances, and indoor temperature, the consumption of electricity and natural gas within each energy stock can be modelled. The forecast of electricity and natural gas consumption represents a significant planning tool for the demand side monitoring of in order to provide a consumption-saving offering of energy supply to clients.The generalized polynomial regression based DSMonitor™ app characterizes the facility of individual consumer in order tocapture consumption as criterion variable, Q_i, (see Equation (9)) for each of the energy commodity.

$$Q_i = a_{i0} + \sum_{j=1}^{m} a_{ij} X_{ij}, i = 1, 2, \cdots, n \tag{9}$$

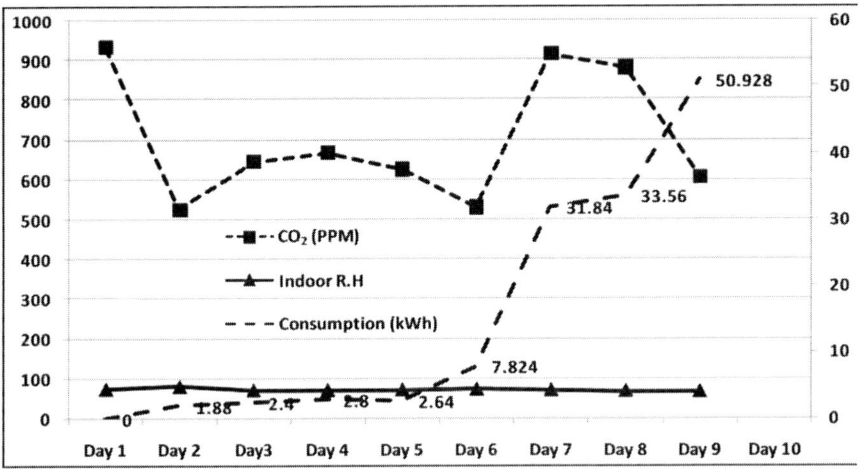

Figure 13. Bottom-up model based on thermal comfort and IAQ for the ELT stock.

where n represents total number of energy commodities, X_{ij} denotes the predictor variables, and the regression coefficients are denoted by a_{ij} The regression coefficients are characterized with the thermal comfort and indoor air quality at the Engineering Lecture Theatre (ELT) of the Faculty of Engineering, University of Lagos (Figure 13).

4. CONCLUSION

Demand side monitoring of residential and commercial facilities in the City of Ottawa, the City of Guelph, and Nigeria is conducted. Characteristics of energy components within each facility, in addition to time-series

modelling and data-mining, are captured. Analysis of these data results in an independent forecasting model for each energy stock, thereby generating an effective tool for engineering service recommendations and designs, which is recommended. In addition, exergetic analysis of the developed energy models led to recommendation of specific iterative service intervention; and path to possible enhanced energy system efficiency. The system efficiency of the proposed high efficiency homes and businesses is projected to lead to economic and environmental saving advantages.

5. ACKNOWLEDGEMENTS

The efforts of Green Energy Associates of Energhx and research students at the University of Lagos are appreciated. Also, the co-operation of Energhx ESCo clients, whose residences and businesses were used for this study, is greatly acknowledged.

REFERENCES

1. E. O. B. Ogedengbe, Z. A. Adem, D. Kraehling and A. Mathur, “Demand Side Monitoring with Solar PV-Based Renewable Energy Supply to Commercial Sectors,” 26th European Photovoltaic Solar Energy Conference and Exhibition, 6CV.1.16, PV Taking-Off: Large Scale Deployment, Hamburg, 5-9 September 2011.

2. H. Lund, A. Marszal and P. Heiselberg, “Zero Energy Buildings and Mismatch Compensation Factors,” Energy and Buildings, Vol. 43, No. 7, 2011, pp. 1646-1654.

3. W. Wang, Y. Xu and M. Khanna, “A Survey on the Communication Architecture in Smart Grid,” Computer Networks, Vol. 55, No. 15, 2011, pp. 3604-3629.

4. N. C. Nair and L. Zhang, “SmartGrid: Future Networks for New Zealand Power Systems Incorporating Distributed Generation,” Energy Policy, Vol. 37, No. 9, 2009, pp. 3418-3427.

5. R. B. Hiremath, B. Kumar, P. Balachandra and N. H. Ravindranath, “Bottom-Up Approach for Decentralised Energy Planning: Case Study of Tumkur District in India,” Energy Policy, Vol. 38, No. 2, 2010, pp. 862-874.

6. P. Denholm and R. M. Margolis, “Evaluating the Limits of Solar Photovoltaics (PV) in Traditional Electric Power Systems,” Energy Policy, Vol. 35, No. 5, 2007, pp. 2852- 2861.

7. A. A. Solomon, D. Faiman and G. Meron, “An EnergyBased Evaluation of the Matching Possibilities of Very Large Photovoltaic Plants to the Electric Grid: Israel as a Case Study,” Energy Policy, Vol. 38, No. 10, 2010, pp. 5457-5468.

8. D. Butler, “Surplus Power Costs Ontarians $35M,” Ottawa Citizen, Monday, A1, 25 July 2011.

9. IEA DSM Task XVI, "Competitive Energy Services (Energy-Contracting, ESCo Services)," Final Task Report, Phase 1, 2006-2009, Graz, 2010.

10. E. O. B. Ogedengbe and G. F. Naterer, "Finite Volume Computations of Convective Exergy Losses in Microfluidic Devices," International Journal of Exergy, Vol. 5, No. 2, 2008, pp. 117-131.

11. E. O. B. Ogedengbe and M. A. Rosen, "Electro-Kinetic Pumping with Slip Irreversibility in Heat Exchange of CSP-Powered Bio-Digester Assemblies," Entropy 2012, Special Issues of Exergy: Analysis and Applications, Vol. 14, No. 12, 2012, pp. 2439-2455.

12. M. Filippi, "Linee di Ricerca per una Procedura Finalizzata alla Certificazione Energetica Degli Edifici," Ingegneri Torino, Technical Review of the Engineer Association of the District of Torino, Vol. 1, 2005, p. 47.

13. M. Caldera, S. P Corgnati and M Filippi, "Energy Demand for Space Heating through a Statistical Approach: Application to Residential Buildings," Energy and Buildings, Vol. 40, 2008, pp. 1972-1983.

14. S. Schiavon and A. K. Melicov, "Energy Saving and Improved Comfort by Increased Air Movement," Energy and Buildings, Vol. 40, No. 10, 2008, pp. 1954-1960.

15. E. O. B. Ogedengbe, "Thermal Management with SolidFluid Slip Irreversibility Treatment in Conjugate Devices," Journal of Thermodynamics, Vol. 2009, 2009, Article ID: 176495.

16. E. O. B Ogedengbe, M. A. Rosen and G. F. Naterer, "Slip-Flow Irreversibility of Dissipative Kinetic and Internal Energy Exchange in Microchannels," Journal of Micromechanics and Microengineering, Vol. 16, No. 10, 2006, pp. 2167-2176.

17. E. O. B. Ogedengbe, R. U. Eteure and M. A. Rosen, "A Modified PMV Model for Indoor Thermal Comfort Analysis: Case Study of a University Cafeteria," The Open Renewable Energy Journal, 2013, under review.

18. E. O. B. Ogedengbe and G. F. Naterer, "Convective Flux Dependence on Upstream Flow Directionality in Finite Volume Computations," Numerical Heat Transfer A, Vol. 51, No. 7, 2007, pp. 617-633.

19. Ontario EBT Working Group, "Electronic Business Transaction Standard Document for Retail Settlement in the Electric Retail Open Access Industry," Version 3, February 2005.

20. Ontario EBT Working Group, "EBT Data Transport Protocol," Version 3, February 2005.

21. P. Sakulpipatsin, L. C. M. Itard, H. J. Van der Kooi, E. C. Boelman and P. G. Luscuere, "An Exergy Application for Analysis of Buildings and HVAC Systems," Energy and Buildings, Vol. 42, No. 1, 2010, pp. 90-99.

22. E. O. B. Ogedengbe, D. Kraehling and Z. A. Adem, "Demand Side Monitoring of Energy Systems in Ontario's Residential and Commercial Sectors," 9th Annual International Energy Conversion Engineering Conference, AIAA 2011-5871, San Diego, 31 July-3 August 2011.

23. J. Widen, A. M. Nilsson and E. Wackelgard, "A Combined Markov-Chain and Bottom-Up Approach to Modelling of Domestic Lighting Demand," Energy and Buildings, Vol. 41, No. 10, 2009, pp. 1001-1012.

24. I. Sartori, B. J. Wachenfeldt and H. G. Hestnes, "Energy Demand in the Norwegian Building Stock: Scenarios of Potential Reduction," Energy Policy, Vol. 37, No. 5, 2009, pp. 1614-1627.

25. S. K. Aggarwal, L. M. Saini and A. Kumar, "Electricity Price Forecasting in Deregulated Markets: A Review and Evaluation," Electrical Power and Energy Systems, Vol. 31, No. 1, 2009, pp. 13-22.

9
CHAPTER
RECENT DEVELOPMENTS ON SILICON BASED SOLAR CELL TECHNOLOGIES AND THEIR INDUSTRIAL APPLICATIONS

Jiahe Chen[1, 2]

[1] College of Materials Science and Engineering, Xiamen University of Technology, Xiamen, Fujian, P.R. China

[2] Free consultant, South Haida, Haicang, Xiamen, Fujian, P.R. China

1. INTRODUCTION

Solar energy is one of the most important alternative energy resources to the traditional fossil fuel energy. Since solar energy merely utilizes the solar radiation lights inducted from the sun, which are endless and available periodically to the earth planet, it is considered a renewable energy. It does not depend on the geographic location of the specific sources and is available everywhere in all countries in the planet. Since solar energy is completely natural, it is also considered a clean energy source. It does not disrupt the environment or create a threat to our eco-systems the way fossil fuel and some other energy sources might. It does not cause greenhouse gases, air or water pollution. With the development of solar technology in the past decades, solar energy becomes an economically affordable energy source and attracts more and more countries to include it in their national strategy for development.

There are several applications utilizing the solar energy, among which the photon thermal utilization and photovoltaic (PV) application are the most common applications. Photon thermal system generally converts the sunlight energy to thermal energy by collecting heats to the medium distributed within a structure or a district heating network. PV system is using the so called photovoltaic effect to convert sunlight energy into electricity. The energy conversion efficiency of a typical photon thermal system is around 28% while a typical PV system is around 18% in the current market. The utilization of photon thermal system normally requires complicate configuration in the system integration and is limited by the thermal consumption. PV system generating electricity can be used for direct residential and industrial utilization whose consumption is unlimited. Both photon thermal system and PV system require the conversion device to be positioned in the sunlight receiving surface, which is limited by the area of rooftop or land. Higher conversion efficiency of solar system is pursued in the market balancing the investment and the return.

In a PV system, the solar cells exercise the photovoltaic effect and determine the conversion efficiency of the whole system. Traditionally, semiconductor materials in the solar cell are doped to form P-N structure as an internal electric field. Among plenty of the semiconductor materials, silicon has a development history up to 50 years. The p-type (positive) silicon has the tendency to give up electrons and acquire holes while the n-type (negative) silicon accepts electrons. When sunlight hits the solar cell, the photons in light excite some of the electrons in silicon to become electron-hole (negative-positive) pairs. Under the internal electric field established in the P-N structure, these pairs are induced to separate. As a consequence, the electrons move to the negative electrode while the holes move to the positive electrode. If a conducting wire connects the negative electrode, the load and the positive electrode in series to form a circuit, an electric current is generated to supply the external load as a result. This is how the PV effect works in a solar cell.

Solar cells are assembled together to form solar modules, which can be arranged into arrays to form a so-called PV system that is large enough to function as a power station for industrial, commercial, and residential use. Through PV effect, the solar cells capture sunlight and turn it into direct current (DC) electricity. For off-grid PV system, the DC can be immediately used for the DC loads or the DC can be directed to an inverter, which converts DC into alternating current (AC) that is suitable for conventional electric appliances. In the off-grid system, the excess energy generated from the PV modules is usually stored in energy storage unit such as capacitors or batteries, controlled by charge controller, for use at night when there is no sunlight or when a larger energy consumption than generation is loaded. An optional backup power, such as diesel generator, can be installed if electricity from the energy storage unit runs out. For grid-tied system, DC electricity generated from the solar cells is converted into AC to be used on-site or stored for backup if the PV system includes energy storage unit. When there is more demand, power can be drawn from the tied grid. Excess supply of electricity from the PV modules can also be fed back into the grid after tuning the phases and frequency. This process of drawing and/or feeding electricity to the tied grid can be monitored by the solar production meter and the export/import meter.

Among all the components in a PV system, the solar cell that converts sunlight to electrical energy is considered to be the most critical device. Solar cell technology is thus deemed to be the most important sector in the solar industry for several decades. Increasing the conversion efficiency and reducing the cost of silicon based solar cells are the two mainstream trends in recent years. This paper attempts to briefly overview the developments in the fields of solar cell silicon materials, the emerging solar cell device architectures and the corresponding immerging silicon based solar cell manufacturing processes, especially from an industrial application point of view.

2. TRADITIONAL SILICON MATERIAL, CELL STRUCTURE AND PROCESSES FOR SILICON BASED SOLAR CELL

Silicon and some other semiconductor materials are the basic materials of solar energy. Today, about 90% of solar cells that installed globally are made from silicon. There are generally two types of this semiconductor: monocrystalline (also called single crystal) silicon and multicrystalline silicon. Multicrystalline silicon is composed of a number of smaller silicon crystals.

Figure 1 shows a traditional manufacturing process from raw silicon (polycrystalline silicon) to final solar modules (also called solar panel) [1]. At first step, silica goes through carbothermic reduction process and forms the metallurgical grade silicon at a purity level of 2N-3N.The metallurgical grade silicon is then refined and subjected to the casting scratch process to become polysilicon raw materials at a purity level of around 6N. For growing the silicon crystals, polysilicon material is then introduced into two different production process for mono-crystalline silicon and multicrystalline silicon production: monocrystalline silicon are generally produced with the Czochralski (CZ) method and multicrystalline silicon are commonly made with directional solidification methods (also called the brick casting method). The monocrystalline silicon ingot is then sliced into wafers. As for the multicrystalline silicon, the silicon brick is first diced into bars and then sliced into wafers [2]. The multiple crystals create boundaries for electrons resulting in less efficiency comparing to monocrystalline silicon. The difference in cell conversion efficiency between these two type silicon solar cell modules is typically 1.5-2% [2]. However, multicrystalline silicon can be produced at a lower cost than the mono-crystalline and it is used most in the solar industry. For traditional silicon based solar cells, the dopant in silicon is p-type (boron).

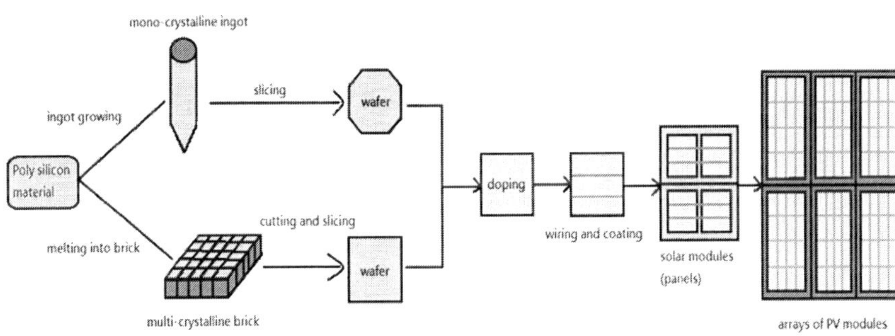

Figure 1. Traditional manufacturing process from raw polysilicon to the final solar module [2]

Compared to monocrystalline silicon, the surface of multicrystalline silicon wafer is more difficult to be passivated due to the existence of grain boundaries and the various grain crystallization orientations, which typically results in about 0.5% conversion efficiency loss. The crystallization defects in the bulk of multicrystalline silicon, such as grain boundaries and metal impurity contaminations, generate the carrier recombination centers and hence degrade the conversion efficiency further [3, 4]. However, ascribed to the low cost of crystallization, multicrystalline silicon wafers have gained more than 40% market share in the global solar market. The monocrystalline silicon wafer is popular especially in the rooftop applications due to its higher cell conversion efficiency per unit covering rooftop area.

Screen printing technology mainstreams the manufacturing processes for traditional silicon based solar cell. Figure 2 shows a structure and fabrication flow for a conventional screen printed solar cell [5]. Generally, the major processes in the common industrial screen printing technology include six core steps: texturing and surface cleaning, phosphorous doping diffusion to generate p-n junction, coating antireflection layer on the front side, printing aluminum paste on rear to form back side field (for metal impurity gettering, light reflection and rear passivation) and silver pastes on both sides, firing the printed pastes to form contacts to front emitter and rear base and isolating the emitter and base on cell edges. There are quite a few review papers covering the detailed description of the screen printing technology [6, 7].

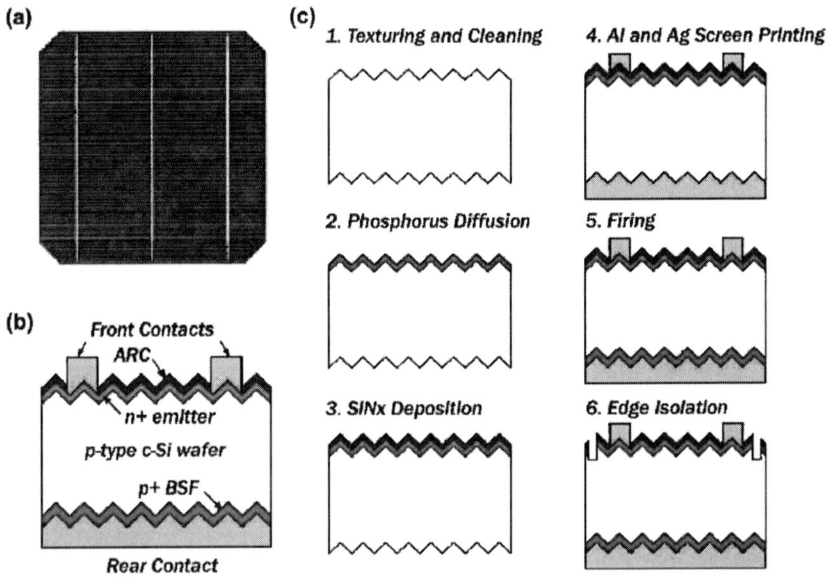

Figure 2. A structure and fabrication flow for a conventional screen printed solar cell. (a) A photo of a typical solar cell on a monocrystalline Si wafer. (b) Schematic diagram of the cross section of screen printed solar cell. (c) six core process flow steps which are used in manufacturing of basic screen printed cells. [5]

3. NEW DEVELOPING SILICON CRYSTALS AND THEIR CHALLENGES TO THE TRADITIONAL SILICON BASED SOLAR CELL PROCESSES

When the solar market demand shifts to the high efficiency side, to low down the preparation cost of material and improve the degradation resulted from the intrinsic and external defects are widely pursued in the silicon manufacturing industry. New silicon materials, such as cast-mono silicon, large-sized grain multicrystalline silicon and ultra-thin crystalline silicon are the representative silicon materials under development. Moreover, the n-type silicon material also shows its growing tendency to be adopted as the solar cell substrate. Additionally, one shall note that recent market demands in the distributing PV system increases the market share of cells made from monocrystalline silicon steadily.

3.1. Quasi-monocrystalline silicon

The directional solidification method (also called casting method) was used to grow single crystals. The directional solidification system (DSS) furnace used for multicrystalline silicon production was originated from the heat exchange method (HEM) furnace for single crystals of sapphire. The growth of single crystal silicon with the directional solidification (or casting) method and the application of the resulting mono (or quasi-mono) silicon wafers for solar cells was also carried out by numerous researchers since 1987, implemented in large scale pilot application since 2006 and attracted the interests from industrial applications since 2011 [8]. Quasi-monocrystalline silicon is casted with the common crystal growth furnaces for multicrystalline silicon. There are two casting approaches for obtaining the quasi-mono-crystalline silicon ingots, one is cast with the pre-grown seeds (cast-on-seed) and the other is to cast the ingot without seeding (seedless casting).

Casting ingot with seeds has been successfully commercialized. The major difference in the cast-on-seed process of growing the quasi-monocrystalline silicon ingots, compared to the growth process of multicrystalline silicon ingots, is that a layer of monocrystalline silicon seeds are employed on the surface of the bottom side of the crucible [9]. These seeds are usually sliced from the monocrystalline silicon ingot grown with the CZ method. Silicon feedstock and dopants are then loaded on top of the seed. In the process of melting, the silicon feedstock is controlled to start the melting from the top so that the seeds will not be completely melted. In the process of solidification, the silicon melt starts to nucleate on the remaining seeds and the grown crystal follows the orientation of the seeds and gradually form a large ingot consisting of crystals with some sections having multiple grains, especially in the areas close to the crucible and between the seeds. As a result, the ingot is with large single crystalline grains present in the center while plenty of small-sized grain exists in the edge regions of the same ingot. Figure 3 shows the image of silicon wafers with different categories of quasi-mono wafer sliced from the same ingot. [8]

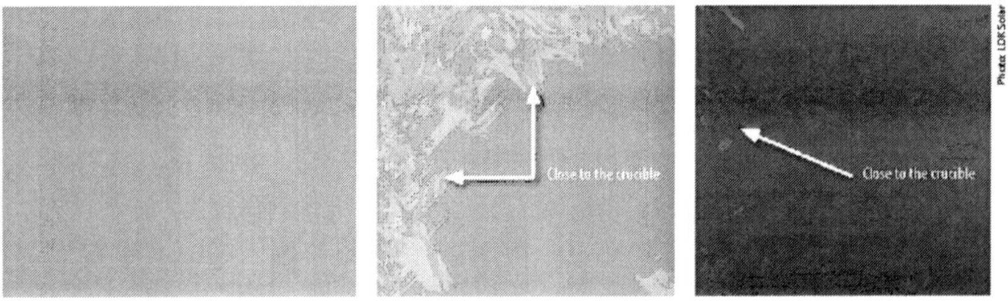

Figure 3. Image of silicon wafers with different categories of quasi-mono wafer sliced from the same ingot [8].

Quasi-mono crystalline silicon can also be obtained without the pre-arranged CZ seeds. However, the challenges rely on a more precise control of the temperature distribution gradient in the silicon liquid as well as the solidification speed of the crystal ingot. In principle, a smoother temperature distribution gradient as well as a slower but steady solidification speed is beneficial for coarsening the grain sizes and in turn reduces the grain boundaries in the casted ingot, which can form the large grain multicrystalline crystal and eventually form quasi-mono crystal if the nucleation sites during solidification process are controlled ideally. Using such approaches to obtain quasi-mono crystal is still under early stage of development whereas using the principle to grow the large grain multicrystalline silicon has been adopted in practice. The electrical behavior of multicrystalline silicon is shown to be influenced by the properties of the grain boundaries, the study of the grain boundaries is very important for improving the efficiency of multicrystalline silicon solar cells over decades. Improving the thermal filed control through some simple retrofits on the traditional casting furnaces can form a multicrytalline silicon ingot with the large grains sizing up to 120-220mm. In such ingot, the amount of grain boundaries are greatly reduced compared to the common multicrystalline silicon ingot whose grain sizes are typically few millimeters [10].

Compared to the CZ monocrystalline, the small angle sub-grain boundaries originating from the gap of the seeds can be observed from the centered quasi-monocrystalline silicon, which increases the recombination of carriers. Nevertheless, the oxygen content of quasi-mono is less than that of mono-crystalline silicon, which reduces the effect of light induced degradation of quasi-mono solar cells. In such a way, the cell efficiency made from quasi-mono wafers with large percentage of quasi-mono grain is comparable to the mono wafers. However, for the quasi-mono wafer with low percentage of quasi-mono grains, the dislocation density, the concentration of oxygen and carbon, and the impurity made the efficiency obviously lower than that of a high percentage of large quasi-mono grains. The properties of the silicon wafers sliced from the edge regions of quasi-mono crystalline ingot are closed to the common multicrystalline silicon wafers. Typically, if the efficiency of a mono silicon cell is 18.5%, the efficiency of a quasi-mono silicon solar cell with a high large single grain percentage can reach about 17.5 to 18.2%. But for low single grain percentage quasi-mono silicon solar cells, the efficiency can be as low as about 16.6 to 17.0%. [8]

The major challenge that quasi-monocrytalline silicon materials brought to the common solar cell processing is mainly the texturing compatibility. In general, for different percentages of large grains of the quasi-mono wafer, different texturing processes should be selected, so as to minimize the textured surface reflectivity and in turn maximize the solar cell conversion efficiency. Typically, the alkali texturing process can be applied to the quasi-mono wafers with mono grain or large percentage of mono grain, so that the light reflection on the solar cell surface can be greatly reduced. For the large quasi-mono grain, alkali texturing can form an inverted pyramid texture surface, similar as the mono crystalline wafers, and improve the efficiency of the solar cell. But due to the anisotropic etching of alkaline, the pyramids that form on the other grains with an orientation different from that of the large quasi-mono grain will have a different structure and orientation. This area has a different reflectivity for the induced light and exhibits different colors from that of the large quasi-mono grain. For a quasi-mono wafer with low percentage of large quasi-mono grains, acid texturing can be applied similar to that for multi silicon wafers [10]. Some compensated texturing processes comprising multiple-step alkaline and acid texturing as well as adopting specific additives in the wet-chemical solutions are also developed [11].

3.2. N type silicon

The very first solar cell, fabricated in 1954 in the Bell Labs, was made of a monocrystalline n-type Si wafer. However, the main industrial application of solar cell was for space applications like satellites until the 1980s. Because the p-type silicon was proved to be less sensitive to the degradation caused by exposure to cosmic rays, such as high-energy particles (protons and electrons), all industrial solar cell development was based on p-type silicon for decades. Since past decade, n-type (mainly phosphorus-doped) silicon material and related cell processes for the territorial application starts to attract a lot of attentions from the scientific researchers.

Compared to standard p-type (boron-doped) silicon solar cells, n-type silicon solar cells feature two important advantages. In one aspect, n-type silicon does not suffer from light induced degradation caused by the simultaneous presence of boron and oxygen (B-O pairs) in the wafers, a phenomenon that in standard p-type silicon solar cells leads to a reduction of the module power output by usually 2-3% within the first few weeks of installation. In another aspect, n-type silicon wafers are less sensitive to metal impurities that are usually present in the silicon feedstock and less effort has to be made to obtain n-type silicon wafers with a high electronic quality. In addition, the performance at low light intensities is predicted to be higher for n-type solar cells, as the lifetime increases in contrary to p-type cells [12], thus enabling an increased power output averaged over the year. As a consequence, n-type silicon wafers featuring high solar cell efficiency potential can be produced more cost effectively than the high quality p-type silicon wafers.

Although the n-type silicon has not yet been studied exhaustively, the industrial applications of n-type silicon have been pushed by the market. Very high quality monocrystalline n-type ingot can be pulled via CZ pulling method and the n-type wafers featuring a high diffusion length of the charge carriers can be manufactured in a routine industrial process used for p-type silicon. However, the challenge of n-type (phosphorus-doped) silicon material is, compared to the standard p-type (boron-doped) silicon, the poor homogeneity of the electrical properties throughout the height of the silicon ingot ascribed to the low segregation coefficient of phosphorus (0.35) compared to boron (0.8) dopant in the silicon liquid during solidification [13]. For instance, for boron doping, a range between 1-3 ohm.cm can be easily maintained throughout the height of whole ingot, while in the case of phosphorus doping, this range increases to 3-12 ohm.cm or more. As standard solar cell concepts require a narrow resistivity distribution to allow stable efficiencies for mass wafers from same ingot, the large variation of the resistivity of n-type silicon ingot decreases the yield for solar cell production thus increasing overall production costs. One solution is to develop new solar cell structures that are less sensitive to the base resistivity; however, a more feasible solution is the application of ingot growth techniques based on the continuous feeding of Si feedstock. Such continuous CZ-pulling technologies can result in a better homogeneous and higher quality electrical properties distribution in the n-type silicon ingot and are currently under development.

The adoption of n-type silicon as the solar cell substrate also brings several challenges to the industrial processes established for traditional p-type solar cell. There are quite a lot of academic and industrial groups working on developing the n type cell technologies in various manufacturing processes, among which the aluminium rear emitter, passivation for p+surface and the metallization are three of the key fields. However, besides the improvements in solar cell technology, it is also necessary to advance the modules, in order not to lose the benefit of the enhanced performance of the cells. The most important topics besides material savings are the improvement or replacement of the absorbing EVA by silicones and the replacement of soldering by gluing techniques which will not be discussed herein.

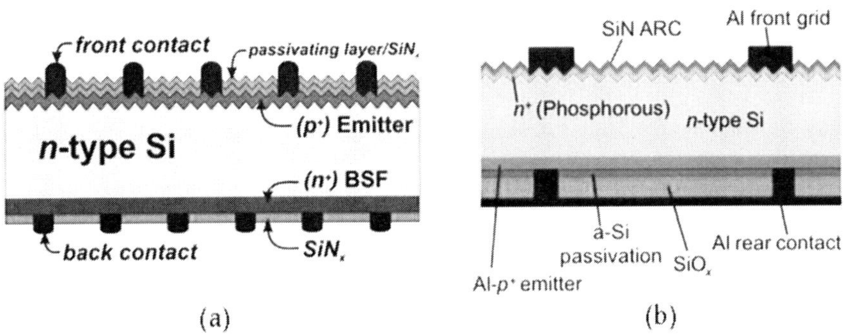

Figure 4. Two schematic cross-section of typical concept of n-type silicon based solar cells. (a) a fabricated n-type solar cells with front emitter [14]. (b) with screen printed rear Al-p+ emitter, full-area metalized a-Si passivated emitter and with local point contacts [15].

The most essential process for good processing of advanced n-type solar cells is the creation of a homogeneous and adapted p+ emitter. Figure 4 shows a typical n-type solar cell. [15] The rear p+ emitter is usually generated by the recrystallization process forming an Al-doped region [16].

In the more advanced solar cell structures, such as HIT concept, a boron-doped emitter is formed by the growth of an amorphous, boron-doped thin layer. Intrinsic and doped amorphous silicon layers are deposited at low temperatures on the front for the emitter (silane and diborane) and on the rear for back surface field creation (silane and phosphine) [17]. In such process, the cleanliness of the surface before amorphous silicon layer depositions is extremely important and several wet chemical processes are developed, for example, Piranha cleaning of H_2O_2/H_2SO_4 and HF/O_3 cleaning. Low temperature (<200°C) screen printing pastes also has to be used for such process since these layers are not stable against high temperatures above 300-400°C. Furthermore, boron-doped emitter for n-type silicon can be also realized by the high temperature tube diffusion process using boron containing dopants as well as by ion implantation followed by thermal recrystallization anneal process [18].

However, the silicon nitride (SiN_x) layer formed by plasma enhanced chemical vapor deposition (PECVD) that widely used for passivating the n+surface of p-type silicon not only has no passivating effect but even deteriorates the surface passivation with respect to an unpassivated surface for boron-doped p+ emitter of a n-type cell. The efficient passivation material for n-type silicon includes, a stack consisted of a thin thermal SiO_2 (formed by anneal processing at a short process time and a low temperature), a PECVD SiNx layer and an aluminum oxide (Al_2O_3) layer deposited either by atomic layer deposition technology or PECVD. A stack of boron silicate glass (BSG) and PECVD SiN_x is also proposed [19]. Such stack has the advantage that no additional equipment is needed as the BSG is already formed during the formation of the boron emitter and the SiN_x requires only standard PECVD equipment as used in p-type cell production lines.

The classical screen printing process can be applied for n-type cell, however, the pastes have to be modified. As low temperature firing is preferred for the n-type cell and p+ surfaces have to be contacted, a simple Ag paste for p-type cell is not suitable. A silver aluminium (AgAl) paste is used and optimized. The aluminium (Al) in the paste allows a good contact resistance, reduces conductivity strongly and limits the open circuit voltage due to penetration into the space charge region, a substitute of Al in the paste is seeking. Additionally, other screen printed products such as diffusion pastes, diffusion barriers, etching and isolation pastes are gaining more and more importance for advanced cell concepts [18]. Furthermore, the tabbing of Al ribbons to interconnect neighbouring solar cells shall be more careful for n-type cell because the devices will be shunted completed if the AgAl emitter contacts contact the n-type base directly [16].

3.3. Kerfless ultra-thin crystalline silicon

In the 1980s, silicon based solar cells are made on crystalline silicon wafers with typically 400 micron thick, where the silicon material cost corresponds to a large fraction of the total cost of a final solar module. The wafers thickness has been significantly decreased from 400 micron to 200 micron while the cell's surface has increased from 100 cm^2 to 240 cm^2 between 1990 and 2006 [20]. Advanced solar cells were fabricated on wafers as thin as 140 um, resulting to efficiencies higher than 20% in 2006 [21]. Current silicon wafer thickness used for traditional silicon based solar cell is in the 150 micron range.

The development work on the ultra-thin silicon substrate was stimulated by the very high price of raw polysilicon materials. In 2008, the cost of silicon wafer accounted for more than 30% of a PV module cost with approximately 60% of the wafer cost coming from ingot growth and wafering. Additionally, the silicon wafer manufacturing consumed the major fraction of energies used to produce PV modules thus in turn increases the energy payback time. However, the price of polysilicon reduced from $400/kg in 2008 to a price as low as $30/kg in 2013 with the rapid expansion of manufacturing capacity primarily in China and Korea since 2010. Therefore, the advantages of kerfless approaches for reducing the silicon material cost were leveled to a big extent while most manufacturers have failed to scale their technologies and compete with competitors who produce the substrates based on traditional ingot casting and wire sawing. Nevertheless, the use of thin substrate for reducing the consumption of silicon is till of interest for manufacturing high efficiency solar cells based on n-type CZ silicon presently because the n-type silicon material can be up to three times more expensive than the p-type material. However, the cost of n-type wafers may drop as the industry already shows a trend to shift towards n-type technology and some developing advanced ingot growing technologies, such as continuous CZ and magnetic CZ, will be available to a larger number of manufacturers in the near future.

Thinning the wafer can be beneficial for two reasons: reduce the consumption of expensive silicon and gain a higher open circuit voltage in the cell. Instead of wire sawing, lifting off thin wafers from a crystalline substrate could potentially reduce crystalline silicon material lost to kerf, which is usually 120-180 micron thick. Epitaxial growth the crystalline silicon could also potentially provide cost reduction since in this approach the substrates are obtained directly from silane bypassing the growth of polysilicon and pulling CZ silicon ingots. The challenge to produce ultra-thin crystalline silicon solar cells consists of three major tasks: the growth of a high quality large-grained crystalline silicon layer on inexpensive substrate, the incorporation of some light trapping scheme to compensate for the weak near-infrared absorption of crystalline silicon and effective passivation of grain boundaries and surfaces.

In the material preparation level, various approaches to manufacturing the kerfless ultra-thin silicon wafers are developed, among which the two main technologies are ion implantation followed by cleaving [22] and epitaxial growth followed by a lift off [23]. Several lifting off process are developed, including epitaxial growth of silicon film on dissolvable CaF_2 to fabricate the textured monocrystalline silicon film [24, 25], epitaxial growth of silicon on a preformed porous silicon layer as the lifting-off buffer layer [26] and the so-called perforated silicon process [27].

In the solar cell level, two critical requirements in achieving high efficiency on the ultra-thin silicon wafer are advance IR light trapping and very good surface passivation [28]. Light trapping in the ultra-thin monocrystalline silicon cells is important because the optical path length of the IR photons may reach the cell substrate several hundreds of microns. These photons may escape the device after multiple internal reflections through the front surface, thus reducing generation current. Very good surface passivation is also important since the contribution of surface recombination in overall recombination losses increases in the ultra-thin silicon cells and the moderately passivated surfaces may pull the efficiency down.

The basic idea of producing kerfless ultrathin crystalline silicon is to eliminate saws, furnaces, and cell finishing equipment as well as slurry and thick wafers. An ion implant machine is also proposed to replace the diffusion equipment in the commercial production chain. However, whether the new process fits seamlessly into the production flow with little impact on the wafer handling, interconnection, packaging and reliability have not been proven at a full production level. Therefore, developing a new wafer technology and integrating it into an existing process chain remains an enormous challenge [29].

4. IMMERGING ARCHITECTURES OF HIGH EFFICIENCY SILICON BASED SOLAR CELLS AND THEIR REALIZATION PROCESSES

Solar cell is an energy conversion device using the photovoltaic effect, which is basically a device separating the electron and hole in a semiconductor material and typically a p-n junction. However, the modern silicon based solar cell is not a device consisting of a single p-n junction as it was born any more. Thanks to the well-established semiconductor manufacturing technology, wet chemical processing, low cost thin film coating, chemical lithography as well as the laser patterning technologies keep attracting the attentions from of the solar industry globally. Surface structure optimization and better surface passivation both on the light induction front side and the rear side are developed in the past years. The deeper understanding of the device physics and specific material properties, solar cell architectures become more and more complicated. PERC/PERL, MWT/EWT, PHASHA, IBC and SHJ/HIT type solar cells structures are under development in the industry, which would increases the solar conversion efficiency gradually in the next few years.

4.1. Front emitter optimization: selective emitter

As it is shown in Figure 2(b), the most commonly used silicon based solar cell architecture is a planar diode structure, where a thin layer of heavily doped silicon (often called emitter, typically n+) is present at the front surface of a moderately doped wafer of the opposite type (often called base, typically p-type). The emitter area is the region that emits or injects most of the charge carriers under dark operation. In the current standard solar cell manufacturing process, the emitter is formed by in-diffusion at high temperature of an n-type dopant (typically phosphorous) into the surface region of a p-type wafer (typically doped with boron). Besides diffusion, the emitter can also be formed by exploiting ion implantation and/or epitaxy technologies.

A good emitter shall efficiently collect the photo-generated carriers in the emitter region, induce a low-loss lateral transport of the majority carriers from the location where they are collected to the nearby metal contacted region and maximize the output voltage with an optimum doping concentration [30]. In industrial practice, the best emitter is a very thin but heavy doped layer. In the very thin emitter, the photon-generated carriers can be collected from the depletion region as well as the moderately doped base underneath the emitter. The extremely high doping concentration at the surface reduces the contract resistance meanwhile act as a sink for metallic impurity gettering [31]. However, the further optimization of the homogeneous emitter approach requires the development of the printable pastes that can contact the emitters with higher sheet resistance which is a challenge to the front silver paste suppliers. A typical doping of 50 ohm/square on the front side of today's industrial type solar cells is therefore a compromise between the emitter performance and sufficiently low contact resistance.

Selective emitter is developed to overcome the above compromise. The selective emitter structure is normally formed by heavily doped the regions underneath the contact grid and by lightly doped in the light illuminated area at the same time. Figure 5 shows a selective emitter that is a heavily doped region placed directly under the metal line [32]. Such selective doping regime leads to a reduced contact resistance and lower Auger-and SRH recombination and therefore results in an improved blue response and a higher open circuit voltage. Several selective emitter technologies have been developed in the past few years. Among which, the etch-back emitter, inline selective emitter, laser doping and doped silicon ink are successfully commercialized in industrial mass production.

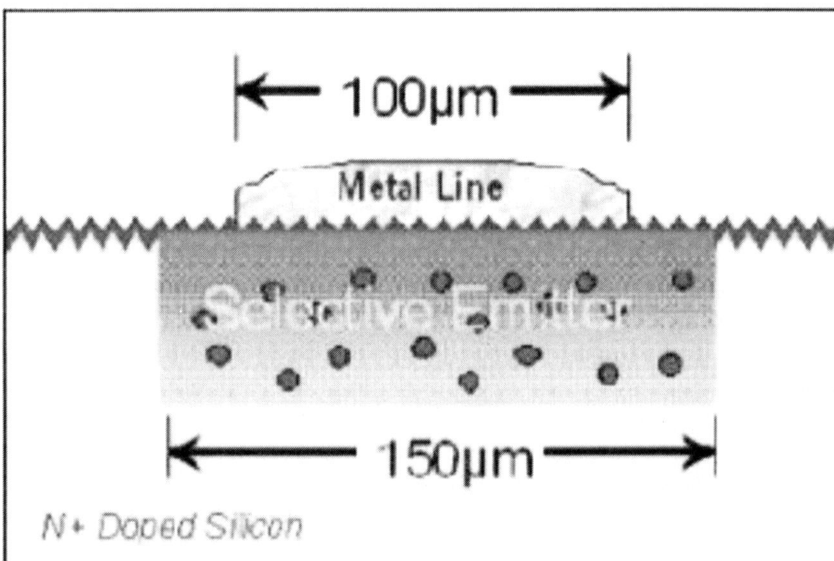

Figure 5. The selective emitter is a heavily-doped region placed directly under the metal line [32].

The etch-back process can be realized with highly homogeneous doping on large area wafers and followed by removing the porous silicon in a wet-chemical solution. The etch-back process in combination with a masking step is an industrially feasible scheme to form a selective emitter structure on p-type wafers. By changing the initial $POCl_3$ diffusion to 20 ohm/square and etching back to 95 ohm/square, a maximum efficiency of a selective emitter solar cell was measured to 19.0% [33].

An inline diffusion process can be realized by coating the wafer surfaces with a defined amount of phosphorus containing dopant with a doper before being laterally transported through a conveyor belt furnace in a controlled ambient at standard pressures. Applying inline selective emitter concept, an increase in open circuit voltage by 18.6 mV and an increase in short circuit current by 1.2 mA/cm^2 were obtained followed by an average efficiency gain of 1.4% and a fill factor improvement by 1.3% compared to homogeneous inline emitters [34].

A laser doping process can be realized by scanning the surface of lightly doped wafers so as to melt the wafer surface locally and enables the fast incorporation of phosphorus atoms from the PSG-layer to form a highly doped selective emitter by recrystallisation. Such process can generate an emitter as thin as 800nm within a few hundred nanoseconds and the emitter will not incorporate any grain boundaries and dislocations. An efficiency gain of 0.5% absolute is reported by adopting such laser doping process [35].

A silicon ink doping process to generate the selective emitter is also commercialised. Such process can be realized by printing the highly doped silicon nano-particles onto the silicon wafer surface prior to the common phosphorous diffusion. The ink is printed only in the areas where the screen-printed front contact is located afterwards. During the following doping diffusion step, a lightly doped emitter (for example 80-100 ohm/square) is realized in the uncovered areas whereas a highly doped areas (for example 30-50 ohm/square) is formed and serves for electrode contacting [36].

4.2. Wafer surface passivation: dielectric materials

Coating a cost-effective antireflection layer is an important step during silicon-based solar cell fabrication. Ideally, the coating should not only reduce optical losses but simultaneously provide a reasonable degree of surface passivation and, for multicrystalline silicon material, a hydrogen passivation of bulk defects and /or grain boundaries. Thermal SiO_2 is one of the obvious candidates for the surface passivation of both n-and p-type silicon of arbitrary doping level and is used in the record-efficiency passivated emitter and the rear locally diffused (PERL) crystalline silicon solar cell [37]. However, the use of a high quality thermal oxide would lead to a high cost of ownership, and the high oxidation temperature may degrade the crystalline silicon bulk quality. Many alternatives based on low-temperature processes therefore receive currently strong research interest, including PECVD, rapid thermal oxidation and electrochemical oxidation [38] Meanwhile, alternative novel dielectric materials, such as SiN_x, SiC_x, SiON and Al_2O_3 are being investigated [39].

SiN_x fabricated by PECVD is increasingly used in industry as it offers the possibility to fabricate a surface and bulk passivating antireflection coating at low temperature (≤450°C) and becomes the most promising candidate for p-type silicon wafers [40]. Recently, eexcellent surface passivation of silicon wafers has been achieved by about 30 nm thick Al_2O_3 layer prepared by plasma-assisted atomic layer deposition (ALD), yielding effective surface recombination velocities of 2 and 13 cm/s on low resistivity n-and p-type silicon, respectively. These results are comparable to the solar cells employing thermal oxide as used in record-efficiency solar cells [39]. Al_2O_3 layer also demonstrated a better UV stability than thermal SiO_2 with the surface passivation improving during UV irradiation [41].

Bi-layer antireflection coatings on the front side have significant advantages over single-layer antireflection coatings due to their broad-range coverage of the solar spectrum. For example, a solar cell with 60 nm / 20 nm SiN_X:H double stack coatings has 17.8% efficiency, while that with a 80 nm SiNX:H single coating has merely 17.2% efficiency. The improvement of the efficiency is due to the effect of better passivation and better antireflection of the stack of antireflection bilayer coating. The double stack antireflection coating is also stable against external stress, which is beneficial for the fabrication of solar module. [42] Bi-layer schemes with different dielectric materials, like SiO_2 capped with SiN_x, have also been explored [43].

Recently, the dielectric rear side passivation of crystalline silicon solar cells has expanded from small lab-scale cells to industrial production. One major advantage of using dielectric passivation instead of a full area Al back surface field is the substantial gain in short circuit current density. A major contribution of this gain stems from the enhanced reflectivity of the rear side of 90-95% at 1000 nm compared to 65% common for fully Al alloyed rear sides in the solar spectrum [44].

Figure 6 depicts the working principle of the solar cell with a rear passivation coating layer [45]. The introduction of backside passivation increases cell efficiency in two ways. Firstly, the backside passivation layer reflects light that has travelled through the cell without generating electrons and reached the backside. The reflected light will pass through the cell a second time generating additional current. Secondly, the layer passivates the crystal matrix defects associated with the back surface of the silicon wafer better than a conventional cell. As a consequence, the electrons generated near the backside of the cell are less likely to be captured and lost. They will have a higher probability of reaching the interface between the base and emitter contributing to the current of the cell improving the cell output voltage. Since blue light will generate more electrons near the front of the cell whereas red light will generate electrons at the back of the cell or even pass the wafer without generating electrons, backside passivation increases the sensitivity of the cell to infrared light with a wavelength of between 1000 and 1180 nm. This additional sensitivity will result in an increased current and in the efficiency of the solar cell.

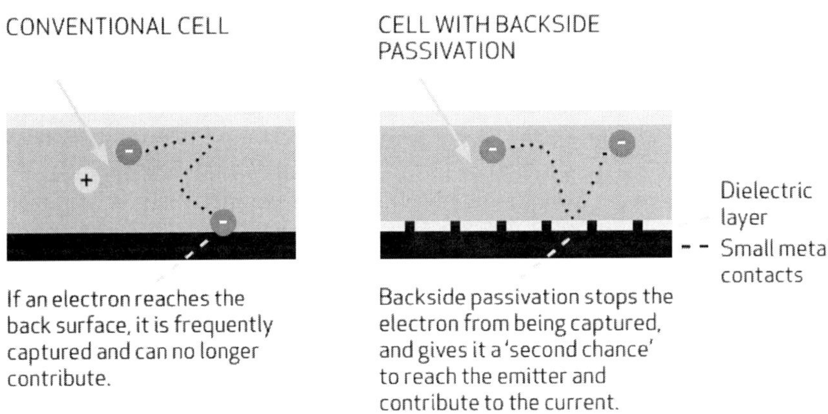

Figure 6. Backside passivation prevents electrons from being captured by the rear surface, resulting in an increased current and voltage of the cell. [45]

Most development efforts are geared towards an evolutionary approach, where the prevailing metallization by means of screen printing and co-firing of thick film paste is retained and the deposition and structuring of the dielectrics is merely added into an already established process [46]. In industrial practise, laser ablation of the dielectric layer for opening the contact channels has been employed on the rear side before printing AgAl paste for contacting the base while some innovative firing through pastes without opening the dielectric layer are also under development.

4.3. Cell structure innovation: Advanced architectures

High efficiency solar cell technologies that can replace screen printed cells in large volume manufacturing in the next 5-10 years are under development. The mainstream high efficiency technologies are passivated emitter and rear cell structure (PERC) or PERL solar cell, metal warp through (MWT) or emitter warp through (EWT) solar cell, bifacial cell with passivated diffusion at the rear (PASHA cell), interdigitated back contact (IBC) solar cell and silicon hetero-junction (SHJ, also known as HIT cells which is a brand name by Panasonic) solar cell.

MWT/EWT technology has been regarded as industrially promising recently because of its high cost-effectiveness for increasing cell and module efficiency [47, 48]. In the MWT/EWT cells, the front metal grids or the front emitter will be wrapped through the laser opened via-holes to the rear side of the wafer inducing reduced shading losses, and reduced surface recombination, and as a result the cell efficiency will be improved [49]. Isolation between emitter and base region is required for MWT cells while the emitter formation on the via-holes is required for EWT cells. Figure 7 shows the cross-section structure of the MWT cells on the top and the EWT cells on the bottom, where the EWT cell is free of contact strips on the front surface [50]. On the MWT/EWT module level, the full back interconnection of the cells results in lower cell-to-module loss thanks to avoiding much of the resistive loss existing in the normal double-side interconnection of H-pattern solar cells with tabs [51]. MWT techniques can be used for commercial solar cells mainly focus on the p-type solar cells as well as n-type materials [52] and is successfully put into mass production.

The structures of the solar cells representing other four main high efficiency technologies are illustrated in Figure 8 [5]. In PERC cells (device a), the efficiency improvement is mainly due to the increase of open circuit voltage which is achieved by replacing a continuous Al BSF with the dielectric passivation layer having the local Al BSF [37]. Additionally, the front surface usually has a higher sheet resistance emitter, with or without higher doping underneath metal contacts (selective emitter), which is usually passivated by thermal SiO_2 and/or AlO_x layers.

One modification of PERC cell design is the PERL cell concept, where the rear local contacts are additionally passivated by p+ boron diffusion. Instead of annealing of the printed or evaporated Al on the rear side, the rear base contact can also be formed by laser firing. Alternatively, the entire rear surface may have additional boron doping formed either by diffusion or implantation which may or may not be passivated by dielectric [5]. Such cell structure concepts are currently incorporated in the conventional p-type cells and have been developing in the campus of a large number of cell manufacturers.

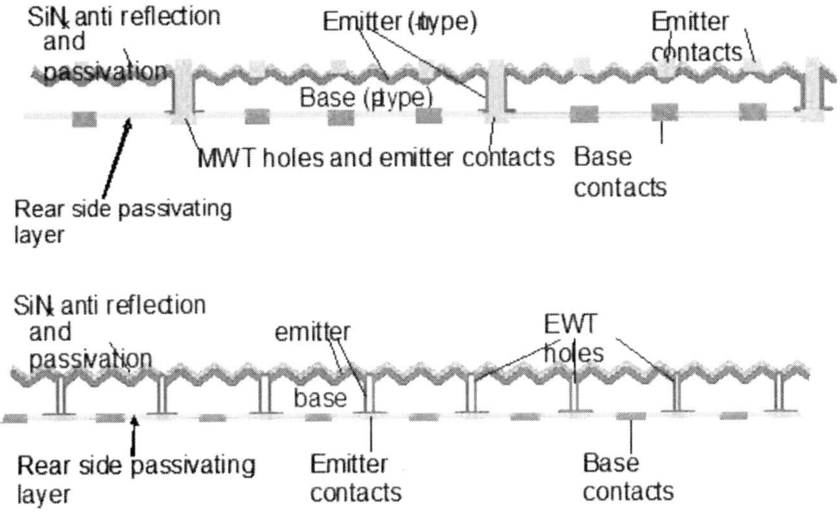

Figure 7. Cross-section structure of the MWT cells (top) and the EWT cells (bottom) [50]

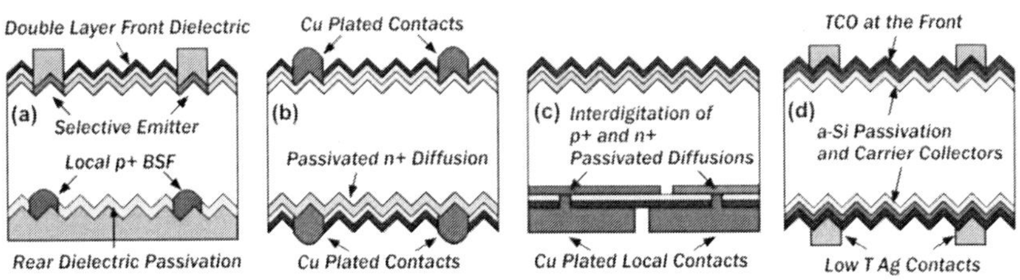

Figure 8. Structure of the solar cells representing four main high efficiency solar cell technologies: (a) PERC/PERL solar cell, (b) bifacial cell with passivated diffusion at the rear (PASHA cell), (c) interdigitated back contact (IBC) solar cell, and (d) silicon heterojunction (SHJ) solar cell. [5]

The other three advanced cell concepts are being developed mainly on the n-type CZ silicon wafers. Bifacial cells (device b) with the passivated diffused carrier collectors and either Al/Ag printed grid or Cu plated grid from both sides of the cell are being developed [53, 54]. The Cu plated metallization increases both short circuit current and filling factor by achieving very thin fingers down to 30 micron width with low resistance, showing the potential of adopting the plating technology in solar industry [55]. IBC cells (device c) utilize the concept that both carrier collectors are formed at the rear side of the cell [56]. This design avoids front metal shading losses and allows the use of thick plated Cu metallization which reduces resistance losses, a recorded efficiency with >21.5% is achieved by the module made from the advanced IBC cells [57]. However, the IBC technology induced a high number of additional manufacturing steps and the use of specialized equipment compared to the common solar cell production line. In SHJ cells, the carrier collectors are formed by depositing thin a-Si layers on both sides of the cell. A transparent conductive oxide (TCO) layer at the front surface (usually Indium Tin Oxide) serves as an antireflection coating, a contacting layer between a-Si and metal electrodes and a lateral conductivity layer. A TCO at the rear surface is usually used for making good ohmic contact and ensuring good internal reflectance of IR light. The efficiency record among large area crystalline silicon solar cells with 24.7% efficiency is reached [58]. The distinguishing feature of SHJ solar cells is due to the almost perfect surface passivation by intrinsic a-Si allowing very high open circuit voltage, which can be higher than 730 mV in commercial solar cell.

5. CONCLUSIONS

Upon reviewing of the recent technology developments on the silicon material, cell device architecture and manufacturing processing for the silicon based solar cell, this paper raises the prospective trends of solar industry driven by the global market demands.

Conclusively, pursuing the highest cost performance ratio of solar cell drives the development of the solar industry. The only measurement of a successful solar cell technology is whether the solar cell can be manufactured in a cost effective way. Restricted by the limitation of area available for installing PV system, the trend of pursing higher conversion energy efficiency does pursed steadily by the terminal users. Therefore, reducing the manufacturing process cost is the only way to survive in the solar industry.

6. ACKNOWLEDGEMENTS

Mrs. Shihua Guo is specially thanked for her contribution on global solar business market investigation, advanced technology evolution tracing and global patent retrieval.

REFERENCES

1. http://pvinsights.com/Knowledge/WhySolar.php
2. Weber, Eicke R. "Transition metals in silicon." Applied Physics A 30.1 (1983): 1-22.
3. Chen Jiahe, EmanueleCornagliotti, Eddy Simoen, EllenHieckmann, Joerg Weber and JefPoortmans, "A deep level transient spectroscopy study on the interface states across grain boundaries in multicrystalline silicon." physica status solidi (RRL)-Rapid Research Letters 5, no. 8 (2011): 277-279.
4. Chen Jiahe, EmanueleCornagliotti, Ellen Hieckmann, Simone Behrendt, Joerg Weber, Eddy Simoen, and JefPoortmans. "On the electrical characterization of grain boundaries in multicrystalline silicon." ECS Transactions 33, no. 17 (2011): 71-80.
5. Herasimenka, StanislauYur'yevich. Large Area Ultrapassivated Silicon Solar Cells Using Heterojunction Carrier Collectors. Arizona State University, 2013.
6. Armin, G. A., Matthew, B. B., Bram, H., & Thomas, M. (2012). Industrial Silicon Wafer Solar Cells–Status and Trends. Green, 2(4), 135-148.
7. Neuhaus, D. H., &Münzer, A. (2008). Industrial silicon wafer solar cells.Advances in OptoElectronics, 2007.
8. http://www.semi.org/sites/semi.org/files/docs/09046%20Standards %20Monolike%20Wafers_jg_PZ.pdf
9. Chinese Patent: CN103320853A
10. Chinese Patent: CN102185028B
11. Chinese Patent: CN101864594A
12. Gong Chun, NielsPosthuma, Frederic Dross, Emmanuel Van Kerschaver, Flamand Giovanni, Guy Beaucarne, JefPoortmans, and R. J. O. M. Hoofman. "Comparison of n-and p-type high efficiency silicon solar cell performance under low illumination conditions." In Photovoltaic Specialists Conference, 2008. PVSC'08. 33rd IEEE, pp. 1-4. IEEE, 2008.
13. Fourmond, Erwann, Maxime Forster, Roland Einhaus, Hubert Lauvray, Jed Kraiem, and Mustapha Lemiti. "Electrical properties of boron, phosphorus and gallium co-doped silicon." Energy Procedia 8 (2011): 349-354.

14. Mihailetchi, Valentin D., Johann Jourdan, Alexander Edler, Radovan Kopecek, Rudolf Harney, Daniel Stichtenoth, Jan Lossen, Tim S. Böscke, and Hans-Joachim Krokoszinski. "Screen printed n-type silicon solar cells for industrial application." In Proceedings of the 25th European Photovoltaic Solar Energy Conference and Exhibition, pp. 6-10. 2010.

15. http://www.isfh.de/institut_solarforschung/alu-zellkonzept.php?_l=1

16. Kopecek, R., A. Halm, L. Popescu, K. Peter, M. A. Vázquez, and Noaki Fukushima. "Aluminium rear emitter large area n-type Cz-Si solar cells for industrial application." In 25rd European Photovoltaic Solar Energy Conference and Exhibition Valencia, Spain. 2010.

17. US patent: US5693957A

18. Radovan Kopecek and JorisLibal, "Switch from p to n", PV magazine, 06/2012, pp.86-94

19. Mihailetchi, Valentin D., Johann Jourdan, Alexander Edler, Radovan Kopecek, Rudolf Harney, Daniel Stichtenoth, Jan Lossen, Tim S. Böscke, and Hans-Joachim Krokoszinski. "Screen printed n-type silicon solar cells for industrial application." In Proceedings of the 25th European Photovoltaic Solar Energy Conference and Exhibition, pp. 6-10. 2010.

20. Photovoltaic Technology Platform; (2007) "A Strategic Research Agenda for PV Energy Technology"; European Communities,2007

21. Mason. N; Schultz. O; Russel.R; Glunz. S.W; Warta. W; (2006) "20.1% Efficient Large Area Cell on 140 micron thin silicon wafer", Proc. 21st EUPVSEC, Dresden 2006, pp.521

22. Henley, Francois, Sien Kang, Zuqin Liu, Lu Tian, Jusong Wang, and Yi-Lei Chow. "Beam-induced wafering technology for kerf-free thin PV manufacturing." In Photovoltaic Specialists Conference (PVSC), 2009 34th IEEE, pp. 001718-001723. IEEE, 2009.

23. Moslehi, M. M. "SolexelInc.,"Thin-silicon, low-cost solar photovoltaic modules using kerfless epitaxial silicon lift-off technology,Standford Seminar on thin-silicon (2012).

24. J. R. Groves, J. B. Li, B. M. Clemens, V. LaSalvia, F. Hasoon, H. M. Branz,C. W. Teplin, "BiaxiallyTextured Photovoltaic Film Crystal Silicon on Ion Beam Assisted Deposition CaF2 Seed Layers on Glass", Energy & Environmental Science. 01/2012; 5(5):6905-6908.

25. Weber, K. J., J. Babaei, P. N. K. Deenapanray, and A. W. Blakers. "Silicon as a Photovoltaic Material." In Materials Forum, vol. 27, pp. 9-14. 2004.

26. PCTPatent Application:WO2012134866A1

27. Brendel, R., H. Artmann, S. Oelting, W. Frey, J. H. Werner, and H. J.Queisser. "Monocrystalline Si waffles for thin solar cells fabricated by the novel perforated-silicon process." Applied Physics A: Materials Science & Processing 67, no. 2 (1998): 151-154.

28. R. Brendel and A. Goetzberger, Thin-Film Crystalline Silicon Solar Cells: Physics and Technology. John Wiley & Sons, 2003.

29. http://www.greentechmedia.com/articles/read/SiGen-Targets-Thin-Silicon-Wafers-with-Ion-Implant-Technology-for-Solar-PV

30. J. Bultman, I. Cesar, B. Geerlings, Y. Komatsu and W. Sinke, "Methods of Emitter Formation for Crystalline Silicon Solar Cells," Photovoltaics International, Petten, 2010.

31. Y. Komatsu, et al., "Innovative Diffusion Processes for Improved Efficiency on Industrial Solar Cells by Doping Profile Manipulation," Proceedings of the24th European Photovoltaic Solar Energy Conference and Exhibition, Hamburg, 21-25 September 2009, pp. 1063-1067.

32. http://www.interpv.net/tech/tech_view.asp?idx=358&part_code=0100 10001&page=3

33. H. Haverkamp, A. Shirazi, B. Raabe, F. Book and G. Hahn, "Minimizing the Electrical Losses on the Front Side: Development a Selective Emitter Process from a Single Diffusion," The33rd IEEE Photovoltaic Solar Energy Conference and Exhibition, San Diego, 11-16 May 2008, pp. 430-433.

34. T. Lauermann, A. Dastgheib-Shirazi, F. Book, B. Raabe, G. Hahn, H. Haverkamp, D.Habermann, C. Demberger and C. Schmid, "Insect: An Inline Selective Emitter Concept with High Efficiencies at Competitive Process Costs Improved with Inkjet Masking Technology," The24th European Photovoltaic Solar Energy Conference and Exhibition, Hamburg, 21-25 September 2009, pp. 1767-1770

35. J. Köhler, P. Grabitz, S. Eisele, T. Röder and J. Werner, "Laser Doped Selective Emitters Yield 0.5% Efficiency Gain," Proceedings of the24th European Photovoltaic Solar Energy Conference and Exhibition, Hamburg, 21-25 September 2009, pp. 1847-1850

36. H. Antoniadis, F. Jiang, W. Shan and Y. Liu, "All Screen Printed Mass Produced Silicon Ink Selective Emitter Solar Cells," Proceedings of the35th IEEE Photovoltaic Solar Energy Conference and Exhibition, Honolulu, 20-25 June 2010

37. ZhaoJianhua, Aihua Wang, Martin A. Green, and Francesca Ferrazza. "19.8% efficient "honeycomb" textured multicrystalline and 24.4%

monocrystalline silicon solar cells." Applied Physics Letters 73, no. 14 (1998): 1991-1993.

38. Chen Jiahe, EmanueleCornagliotti, XaviorLoozen, EllenSimoen, Jan Vanhellemont, Johan Lauwaert, HenkVrielinck, and JefPoortmans. "Impact of firing on surface passivation of p-Si by SiO2/Al and SiO2/SiNx/Al stacks" ,Journal of Applied Physics 110, no. 12 (2011): 126101.

39. Hoex Bram, S. B. S. Heil, Erik Langereis, M. C. M. Van de Sanden, and W. M. M. Kessels. "Ultralow surface recombination of c-Si substrates passivated by plasma-assisted atomic layer deposited Al2O3." Applied Physics Letters 89, no. 4 (2006): 042112.

40. Aberle, Armin G. "Overview on SiN surface passivation of crystalline silicon solar cells." Solar Energy materials and solar cells 65.1 (2001): 239-248.

41. Dingemans, Gijs, et al. "Stability of Al 2 O 3 and Al 2 O 3/a-SiN x: H stacks for surface passivation of crystalline silicon." Journal of Applied Physics 106.11 (2009): 114907-114907.

42. Lee, Youngseok, et al. "Stability of SiNX/SiNX double stack antireflection coating for single crystalline silicon solar cells." Nanoscale research letters 7.1 (2012): 1-6

43. Simoen, Eddy, Chun Gong, N. E. Posthuma, Emmanuel Van Kerschaver, JozefPoortmans, and R. Mertens. "A DLTS study of SiO2 and SiO2/SiNx surface passivation of silicon." Journal of The Electrochemical Society 158, no. 6 (2011): H612-H617.

44. T. Lauermann, T. Lüder, S. Scholz, G. Hahn, B. Terheiden, "Enabling dielectric rear side passivation for industrial mass production by developing lean printing-based solar cell processes", Proc. 35th IEEE PVSC, 2010, p. 28

45. http://www.recgroup.com/Documents/REC_White_Paper_Backside_Passivation_20120713.pdf

46. S. Gatz, H. Hannebauer, R. Hesse, F. Werner, A. Schmidt, T Dullweber, J. Schmidt, K. Bothe, R. Brendel, "19.4%-efficient large-area fully screen-printed silicon solar cells" Phys. Status Solidi RRL, vol. 5, No. 4, 2011, p. 147

47. Clement, Florian, Michael Menkoe, Denis Erath, Tim Kubera, René Hoenig, Wolfram Kwapil, Winfried Wolke, Daniel Biro, and Ralf Preu. "High throughput via-metallization technique for multi-crystalline metal wrap through (MWT) silicon solar cells exceeding 16%

efficiency." Solar Energy Materials and Solar Cells 94, no. 1 (2010): 51-56.

48. Benjamin, Thaidigsmann, Linse Michael, Wolf Andreas, Clement Florian, Biro Daniel, and Preu Ralf. "The path to industrial production of highly efficient metal wrap through silicon solar cells." Green 2, no. 4 (2012): 171-176.

49. E. Van Kerschaver, R. Einhaus, J. Szlufcik, J. Nijs, R. Mertens, "A Novel Silicon Solar Cell Structure with BothExternal Polarity Contacts on the Back Surface", Proceedings of the 2nd World Conference onPhotovoltaic Energy Conversion, Vienna, Austria, 1998, pp. 1479-1482

50. https://www.ecn.nl/news/newsletter-en/2010/march-2010/plasma-for-solar-cells/

51. N. Guillevin, B.J.B. Heurtault, I. Bennett, M.G. Guichoux, L.J. Geerligs, A.W. Weeber, J. Xiong, Z. Hu, G. Li,W. Zhao, Y. Chen, J. Wang, Z.. Wang, C. Chen. "Development towards 20% Efficient n-Type Si MWT Solar Cells for Low-Cost Industrial Production", 26th EUPVSEC, Hamburg, 2011, pp. 989-994

52. WenchaoZhao, Jianming Wang, YanlongShen,Ziqian Wang,Yingle Chen,Zhiyan Hu,Gaofei Li,Jianhui Chen,JingfengXiong,Guillevin,, N,Heurtault B.J.B,Geerligs L.J.,Weeber, A.W., Bultman, J.H.,"0.35% Absolute efficiency gain of bifacial N-type Si Solar cells by industrial metal wrap through technology". In Photovoltaic Specialists Conference (PVSC), 2012 38th IEEE (pp. 002289-002291). IEEE.

53. Romijn, I. G., BB van Aken, J. Anker, A. R. Burgers, A. Gutjahr, B. Heurtault, M. Koppes, E. Kossen, M. Lamers, D.S. Saynova, C.J.J. Tool, Lang Fang, XiongJingfeng, Li Gaofei, XuZhuo, Wang Hongfang, Hu Zhiyan, P.R. Venema and A.H.G. Vlooswijk. "Industrial implementation of efficiency improvements in n-type solar cells and modules." In Proceedings of 27th European Photovoltaic Solar Energy Conference, p. 533. 2012.

54. Hoerteis, Matthias and Stefan W. Glunz. "Fine line printed silicon solar cells exceeding 20% efficiency." Progress in Photovoltaics: Research and Applications 16, no. 7 (2008): 555-560.

55. Lennon, Alison, Yu Yao, and Stuart Wenham. "Evolution of metal plating for silicon solar cell metallisation." Progress in Photovoltaics: Research and Applications 21, no. 7 (2013): 1454-1468.

56. M. D. Lammert and R. J. Schwartz, The interdigitated back contact solar cell: A silicon solar cell for use in concentrated sunlight, IEEE Transactions on Electron Devices, vol. 24, no. 4, pp. 337–342, 1977.

57. D. D. Smith, P. J. Cousins, A. Masad, A. Waldhauer, S. Westerberg, M. Johnson, X. Tu, T. Dennis, G. Harley, G. Solomon, S. Rim, M. Shepherd, S. Harrington, M. Defensor, A. Leygo, P. Tomada, J. Wu, T. Pass, L. Ann, L. Smith, N. Bergstrom, C. Nicdao, P. Tipones, and D. Vicente, Generation III high efficiency lower cost technology: Transition to full scale manufacturing, in 2012 38th IEEE Photovoltaic Specialists Conference (PVSC), 2012, pp. 001594–001597.

58. Taguchi, Mikio, Ayumu Yano, Satoshi Tohoda, Kenta Matsuyama, Yuya Nakamura, Takeshi Nishiwaki, Kazunori Fujita, and Eiji Maruyama. "24.7% record efficiency HIT solar cell on thin silicon wafer." (2013): 1-4.

10

CHAPTER

TRENDS AND POTENTIALS OF THE SMART GRID INFRASTRUCTURE: FROM ICT SUB-SYSTEM TO SDN-ENABLED SMART GRID ARCHITECTURE

Jaebeom Kim [1,2,†], Fethi Filali [1,†] and Young-Bae Ko [2,*]

[1] *Qatar Mobility Innovations Center, Qatar Science and Technology Park, Doha 210531, Qatar*

[2] *Graduate School of Computer Engineering, Ajou University, 206 World Cup-ro, Suwon 16499, Korea*

ABSTRACT

Context and situational awareness are key features and trends of the smart grid and enable adaptable, flexible and extendable smart grid services. However, the traditional hardware-dependent communication infrastructure is not designed to identify the flow and context of data, and it focuses only on packet forwarding using a pre-defined network configuration profile. Thus, the current network infrastructure may not dynamically adapt the various business models and services of the smart grid system. To solve this problem, software-defined networking (SDN) is being considered in the smart grid, but the design, architecture and system model need to be optimized for the smart grid environment. In this paper, we investigate the state-of-the-art smart grid information subsystem, communication infrastructure and its emerging trends and potentials, called an SDN-enabled smart grid. We present an abstract business model, candidate SDN applications and common architecture of the SDN-enabled smart grid. Further, we compare recent studies into the SDN-enabled smart grid depending on its service functionalities, and we describe further challenges of the SDN-enabled smart grid network infrastructure.

Keywords: software-defined networking; smart grid network; Advanced Metering Infrastructure (AMI); context awareness

1. INTRODUCTION

A smart grid is an extensive combination of power grids, communication networks and information management systems, which contributes to green and cost-effective energy generation. Generally, a smart grid comprises a data communications infrastructure integrated with an electrical grid that collects and analyzes data captured in near real time about power transmission, distribution and consumption [1]. Thus, the key to enabling a smart grid system is the seamless convergence of existing electronic grid systems and information-communication technologies (ICTs).

The electrical power grid system can be divided into four elements: electricity generation plants, transmission substations, distribution substations and end users [2]. First, power plants generate power from a variety of sources, including solar, wind and nuclear sources, for distribution. As the power approaches the customers' homes, it is again stepped down to the voltage required for residential use. Finally, home appliances access power through their electric smart meters [2,3,4]. Even though the primary goal of these elements differs, time-critical information that is generated by each element needs to be robustly shared between all elements in a smart grid system. For example, power generation can be dynamically controlled by using the real-time energy consumption of the end users. Meanwhile, the end user can visualize the real-time power usage of the home and can obtain the real-time cost of the power supplied from the power provider. In a smart grid, electricity can also be returned to the grid by users. For example, home users may be able to generate electricity using photovoltaic (PV) systems and return it to the grid or electric vehicles may provide power to help balance loads by "peak shaving", i.e., sending power back to the grid when the demand is high. This backward flow is important. For example, it can be extremely useful in a microgrid that has been "isolated" because of power failures. With the help of the energy feedback from the customers, the microgrid can continue to function, albeit at a reduced level.

By exchanging information between different elements, the smart grid provides predictive information and recommendations to utilities, their suppliers and their customers on how to manage power in an optimal manner [4]. In addition, this information can be collected at the data server and reproduced by the service provider in order to develop new services and business models. Potential smart grid services focus not only on power-related services, but they also aim to enable the development of utility services and business models, such as water, gas, facility monitoring and unmanned surveillance [5,6]. Further, all electric grid elements must share the common understanding that the smart grid needs to be integrated with a data-networking infrastructure in order to be categorized them as "smart things" [7].

In this paper, we revisit the current smart grid system architecture, and we investigate a new paradigm called SDN-enabled smart grid. In addition, we compare recent studies of the SDN-enabled smart grid based on its service functionalities, and we present further research issues regarding the SDN-enabled smart grid. To the best of our knowledge, this is the first work that deals with the identification and comparison of SDN-enabled smart grid architectures.

The rest of this paper is organized as follows. In Section 2, we describe the essentials of the smart grid technologies. In Section 3, we describe the isolated subsystems of the current smart grid that are based on information management and communication layers, namely the information characteristics, communication architecture and its requirements. We present the survey and common design of the SDN-enabled smart grid in Section 4 and Section 5, as well as its challenges and guidelines for the smart grid. In Section 6, we conclude the paper.

2. SMART GRID TECHNOLOGY ESSENTIALS

The smart grid is a large-scale and complex networking system, and the National Institute of Standards and Technology (NIST) has therefore developed a conceptual hierarchical smart grid network infrastructure [1].

This conceptual architectural reference model enables us to analyze use cases, to identify interfaces for which interoperability standards are required and to facilitate the development of a cyber-security strategy [1,9]. Figure 1a shows an example of a three-tier hierarchical smart grid network that is organized by wide area networks (WANs), neighbor area networks (NANs) and home area networks (HANs). The HAN is an end-user network that is constructed by the connection between smart household electric appliances. The NAN is a capillary of the smart grid that provides large-scale data communication and connectivity between each household and a WAN. The WAN is a high capacity data network between NAN gateways and the head-end system (HES). In this smart grid network, all devices need to use the same communication policies, such as the routing strategy and quality of service (QoS) properties of the application, in order to provide the 3S autonomous network functions, namely self-configuration, self-healing and self-optimization. Figure 1b shows the layer-based information and power-flow model of the smart grid. In this figure, all objectives between the power facility and end-user system, such as the power generator, substation and AMI, are connected to the wired power grid infrastructure. The information data from each objective is exchanged via the data-communication infrastructure, which can be established using both wired and wireless communication protocols, such as powerline communication (PLC), Ethernet, IEEE 802.11 wireless local area network (WLAN) and the IEEE 802.15.4 family of protocols [8,9,10].

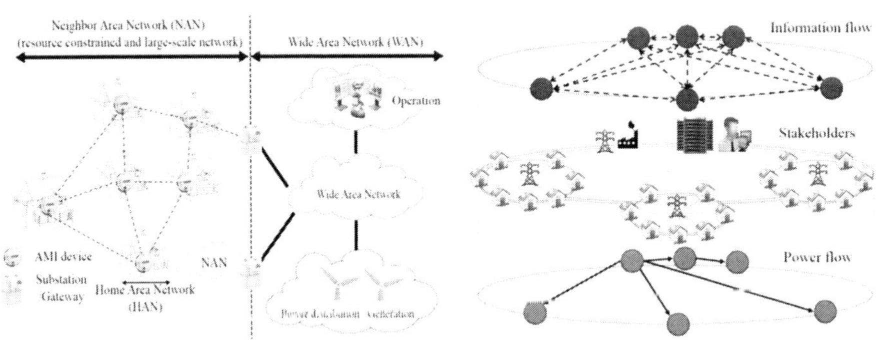

(a) (b)

Figure 1. (**a**) multi-tier-based smart grid network architecture. (**b**) Abstraction of the information and power flow view of the smart grid infrastructure;

In this layered smart grid model, the electric and communication infrastructure support the two-way flow of electricity and information. Note that it is straightforward to understand the concept of the "two-way flow of information". "Two-way flow of electricity" implies that the electric energy delivery is no longer unidirectional. The synthesized requirements of the desired smart grid are as follows:

- The smart information subsystem is responsible for advanced information metering, monitoring and management in the context of the smart grid.
- The smart communication subsystem is responsible for communication connectivity and information transmission between systems, devices and applications in the context of the smart grid.

The smart grid technologies employed for industrial and mission-critical environments require new network-management technologies to provide simple system management between smart grid sub-layers from the application and information sub-systems to the communication infrastructure. Currently, the smart grid concept is a relevant critical-use case, which includes the protection, automation and control of electric power systems and which is supported by ICT facilities that must meet the real-time requirements of these applications with high dependability and security [11].

However, the current communication infrastructure may not be able to dynamically adapt to the new business model of the smart grid because of hardware dependencies. According to some smart grid researchers [12,13,14,15], communication technologies first need to be further developed, and they predict that software-defined networking (SDN) will allow information exchange in different network platforms, different layers and provide the adaption of new business models of the smart grid without hardware replacement. In addition, they argue that the SDN-based smart grid will improve the system robustness by enhancing the reliability and optimization of the transmission asset utilization [13,15]. In the SDN architecture, the control plane of the network is controlled by using a centralized SDN controller, and a service provider can identify not only the network status, but can also flexibly make a network plan for services and business models. This separate control and data plane model of the SDN enables us to find a suitable solution, even if there are potential problems, such as route disconnection, bottlenecks, collisions and application errors without special network devices [16]. However, current SDN is designed for highly capable network environments and needs to solve various challenges to enable the SDN-enabled smart grid in terms of the control message overhead, route construction between devices and interoperability between smart grid protocols.

3. ICT SUB-SYSTEMS OF THE SMART GRID

The smart grid is a complex and interoperable system that needs to process meaningful and actionable information between different subsystems. The exchanged information will be shared by the systems, and this information will elicit agreed-upon types of responses. The reliability, fidelity and security of the information exchanges between smart grid systems must achieve the required performance levels [17]. The ICT technologies are basic building blocks for the underlying data-centric smart grid system. The combined requirements of the desired smart grid ICT are as follows:

- Integrity: The smart grid scopes (generation, transmission, distribution, consumption and control center [1]) and sub-scopes will use a variety of communication networks that are integrated with the IT networks.
- Interoperability: The smart grid will have the capability of two or more networks, systems, devices, applications or components with respect to the ability to exchange and readily use information securely, effectively and with little or no inconvenience to the user [17].
- Interactivity: Utilities and customers will reduce their usage during peak hours. Mechanisms will also be incorporated for consumers to smartly use their power devices to lower their cost.

3.1. Smart Grid applications

From an application perspective, there are three major domains for ICT integration into the power grid [4], as shown in Figure 2. Supervisory control and data acquisition (SCADA) is already widely used for remote monitoring and controlling of higher levels of the distribution grid [5]. This area is also well covered by advanced standards, such as IEC 61850 for substation automation. The active distribution grids are the core of modern smart grids and enable the integration of distributed energy sources mainly into the mid-level voltage grid. The primary task of communication in this respect is to collect critical network parameters and to actively influence the distributed energy generation, so as to maintain the overall power quality. On the lower grid levels, smart metering is the most relevant use case for ICT. Originally intended to enable the remote retrieval of meter values, smart metering also includes load profiling, local power-quality monitoring and the remote switching of loads [23]. This requires a permanent connection to a communication network, even though real-time requirements are not as stringent as in SCADA control applications, e.g., the overall load balancing on the grid.

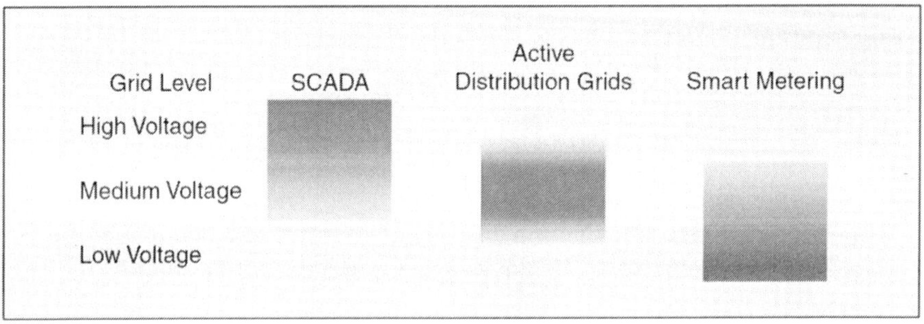

Figure 2. Major ICT application domains in smart grids and their correlation with the grid levels. SCADA, supervisory control and data acquisition.

3.2. Information Management

The traffic volume of smart grid systems has grown rapidly because of extended business models and service domains, such as gas, water and facility surveillance [6]. Although it was an offshoot of recent power grid systems, the smart grid is required to meet more requirements and has to attain new features. This is because the volume of the generated and consumed data traffic is extremely high when compared to the traditional power grid network.

Smart grid needs to use real-time analytic engines that enable the analysis of the power network, determine the current state and condition of the power system, predict what may happen and develop an action plan [24,25]. These engines receive data from the utility and external entities, such as weather services, through an advanced metering infrastructure (AMI). AMI is usually a combination of smart meters, data management modules, communication network modules and applications that are specific to metering and plays a key role in smart grid technology, enabling the enhancement of energy efficiency. Utilities, regulators and policymakers should focus on the consumer-centric architectures for appliance control, the public architecture on which AMI communication is based and the collaborative architecture for content. However, it is evident that choices made by utilities as part of their smart grid deployment strategy and network architecture, which are guided by the policies dictated by state and federal regulatory authorities, may significantly affect the outcome of the smart grid initiatives.

- Control and management: This is intended to monitor and manage all of the components of the electric power system. For example, their behaviors and performances can be modified and predicted to avoid or address potential emergencies [16,18,19,20,33]. In the smart grid system, the HES needs to provide dynamic pricing and end-user device management. Meanwhile, the HES needs to find the optimal configuration parameters of the smart grid system, such as the monitoring pattern, business model, network monitoring and network system configuration [6].
- Neighbor area optimization: This is designed to help customers know the real-time prices of power, enabling them to optimize the power usage accordingly [9,17,21,22]. In addition, consumers then become informed participants and can choose different purchasing patterns based on their needs and the Grid's demand, which can ensure the reliability of the electricity supply [5]. Meanwhile, the communication protocol needs to be optimized to provide stable data delivery services in the resource-constrained environment of the NAN [6].

To realize a stable smart grid, there is a need for information analysis, management and different architectural directions for the smart grid that compares utility-controlled and consumer-controlled energy networks. For instance, with utility control, the intelligence of devices is derived from a central control point via a private utility network. However, with consumer control, these devices use a control system that is located in the home, business or on the Internet, but is ultimately managed by the needs of the consumer [26]. For both of these extremes, in addition to intermediate possibilities, it is important to consider how choices affect the combined efficiency and demand-response application (i.e., energy monitoring), as well as support/limit innovation in supporting a responsive energy future. Many utilities have begun smart grid implementation with AMI.

3.3. Communication Infrastructure

The smart grid is a combination of ICTs and the electrical power grid. It is expected to affect all segments of the current electrical grid system, including generation, transmission and distribution. However, these segments will not function well without an effective data communication networking system. In other words, designing a communication architecture that meets the requirements of the power grid is key to the successful implementation of the smart grid, because power grid implementation has long been established. Based on the requirements of smart grid, it initially implies a need for bidirectional, real-time communication networks for data collection and processing [17]. From a communication technology point of view, information exchange in the upper grid levels for SCADA applications is usually covered by existing communication networks belonging to utilities or grid operation [3].

A smart grid communication infrastructure also includes HANs, building area networks (BANs), industrial area networks (IANs) and a WAN. A HAN is a communication network comprising appliances and devices within a home. A NAN is a network of multiple HANs that delivers the metering data to data concentrators and delivers control data to HANs. A WAN is the largest network for communications to/from data control centers. In a HAN, appliances (such as meters for electricity, gas, water, heat and solar panels) can be equipped with smart meters; these meters also connect smart appliances (i.e., appliances with communications and remote control functions, such as smart dishwashers, dryers and ovens). In addition, these meters connect to a metering gateway. In a NAN, many metering gateways of HANs are connected to each other to form a wireless mesh network [6]. A WAN connects smart metering gateways with utilities and the distribution control system.

These networks are often based on fiber-optic links that are installed in parallel to the high-voltage grid and connect primary substations, large power plants and control rooms. Furthermore, they are mostly based on Internet technologies. Today, the challenge regarding interconnection lies in the medium- and low-voltage levels required, which are usually not reachable (or only to a small extent) by utility-owned communication infrastructures. To bridge this gap, several communication protocols are provided. PLC systems overcome traditional electronic communication technologies. The PLC channel is problematic, because it is time varying and depends greatly on the actual network topology and the connected electrical loads. To ensure connectivity from a communication viewpoint, network elements, such as transformers, need to be bridged with phases coupled, which requires additional equipment that is expensive at the higher voltage levels. Furthermore, regulation of the frequency bands used for data transmission is fragmented worldwide and remains unclear. The implementation of PLC is considered to be both economically and technically challenging and is still in its infancy compared to other communication technologies. However, it is a promising possibility that is currently being investigated.

The use of wireless networks is an alternate option, because they are flexible and can be used to build utility-owned networks. However, typical broadband computer network technologies, such as worldwide interoperability for microwave access (WiMAX) and WLANs, are problematic with respect to their limited transmission range and reachability, especially in urban areas. Hence, they may serve only as communication subsystems for small clusters of devices, such as meters. For long-distance transmission, trunked radio systems, such as terrestrial trunked radio (TTR) or trans-European trunked radio (TETRA), are becoming popular in some countries. Installation costs are moderate, even though the frequencies used are not license free. As TETRA is limited in bandwidth (7.2 kb/s per time slot), it can be considered as a substitute for services that are currently run over GSM/ General Packet Radio Service. Typical applications are metering (load profile meters and households), supervision of transformer stations and safety-related applications [3].

4. EVOLUTION OF THE SMART GRID INFRASTRUCTURE: SDN-ENABLED SMART GRID

The smart grid technologies employed for industrial and mission-critical environments require new network management technologies to provide simple system management between smart grid sub-layers from applications, information sub-systems and communication infrastructure. Currently, the smart grid concept is a relevant critical use case that includes the protection, automation and control of electric power systems and which is supported by ICT facilities that must meet the real-time requirements of these applications with high dependability and security. However, the current communication infrastructure may not dynamically adapt to the new business model of the smart grid due to its hardware dependability. For example, in the traditional network paradigm, each network device, such as switches (AMI meter), individually manages its routing table by using a pre-defined routing protocol. In this architecture, the pre-defined routing protocol of the network device cannot be flexibly changed or configured to adapt network dynamicity. From a service (application) perspective, the system administrator cannot dynamically manage the network system because the network and service layer are isolated from each other. Thus, the service administrator would not be able to directly identify a network problem even when the application service is down due to network instability. Further, a new application profile of the utility service or network function needs to be added to legacy AMI infrastructure, and all AMI devices need to be replaced or upgraded with service interruption. According to smart grid researchers, smart grid communication technologies are required to evolve in order to solve these various complexity problems affecting the large-scale smart grid system. In addition, they predict that SDN will allow information exchange across different network platforms and different layers and will enable the adaption of a new business model of the smart grid system [27,28]. Meanwhile, SDN-enabled smart grid architecture has been considered for various companies, such as Fujitsu and Cisco [46].

SDN is a next-generation network architecture that provides simplified and flexible network-management services through the abstraction of lower-level functionality, such as packet filtering, switching and monitoring. In SDN, to isolate the control and data plane of the network, a centralized network controller manages the control plane of the network instead of individual network switches. The incoming packet information (flow) of each switch is reported to the SDN controller when a packet is not listed in its flow table. The SDN controller decides and publishes the flow entry of the requested flow from the network switches using an SDN application that can be directly programmed by the system administrator. This separated control is implemented in the application layer through an application-network interconnection interface, such as OpenFlow API [29]. Thus, the service administrator manages and monitors not only the service (business) application status, but also the communication system without the need for an additional network management system. Moreover, each resource-constrained AMI device does not need to maintain its routing strategy, which is maintained using the SDN controller. The network design and potential advantages of the SDN-enabled smart grid are discussed below.

4.1. Network Design of the SDN-Enabled Smart Grid System

The SDN-enabled smart grid system can be simply installed over a legacy IP network infrastructure without the need for any additional wired or network installation. This is because the SDN functions are implemented by application software with a general operating system, such as Linux, instead of traditional hardware-dependable network systems. Figure 3 shows an example of an SDN-enabled smart grid. In this example case study, the message exchange between HES and the edge AMI meter is realized using a switch and gateway intelligent electronic device (IED). This communication architecture is the same as the traditional smart grid communication infrastructure, and it can be simply enhanced to become an SDN-enabled smart grid infrastructure by using centralized SDN controllers [4]. In addition, the resource and service capability of the SDN controller is determined depending on its service domain (tier) as a smart grid WAN and HAN. The service function as a communication policy and smart grid (power system) application can be realized within the SDN controller. Meanwhile, the service functions between communication infrastructure and the power grid system application can share the information to optimize the smart grid system.

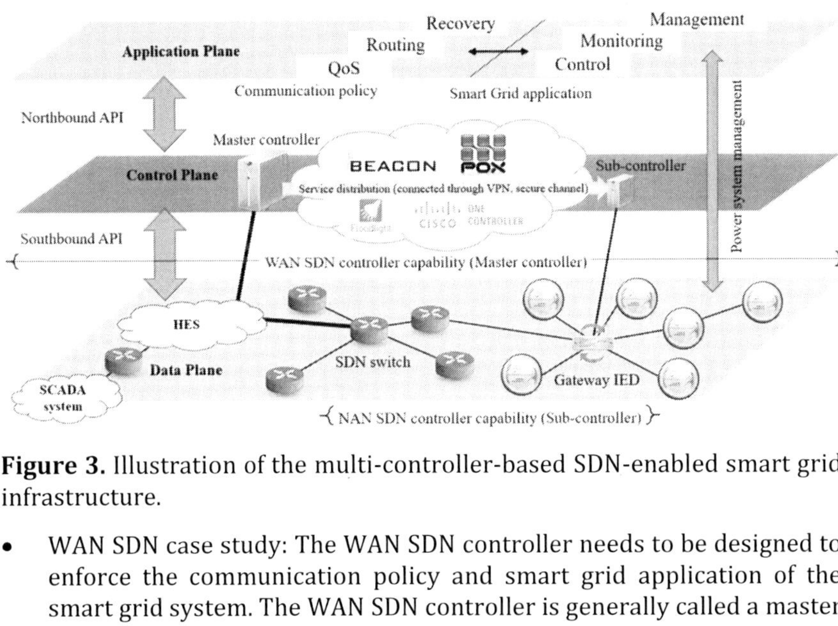

Figure 3. Illustration of the multi-controller-based SDN-enabled smart grid infrastructure.

- WAN SDN case study: The WAN SDN controller needs to be designed to enforce the communication policy and smart grid application of the smart grid system. The WAN SDN controller is generally called a master controller and handles the back-end HES service applications, such as SCADA and intrusion detection system (IDS) control. In addition, its service capability can be extended to micro-grid service applications, such as AMI monitoring systems, depending on the network scale and system resources. Dong, X. et al. proposed a WAN SDN controller architecture that is designed for the back-end SCADA management infrastructure of the smart grid [18,44]. In this system, the master SCADA controller is implemented for one of the SDN applications and controls slave SCADA systems in the power grid network. This WAN SDN controller system handles the global optimization of the smart grid system, such as end-to-end routing, network capacity planning and back-end network management, but it also needs to consider the system resources, such as storage, link capacity and computational power.

- NAN case study: The NAN SDN controller (sub-controller) provides local optimization of the resource-constrained smart grid network. The service capability of the sub-controller is generally limited to a single subnet of the smart grid network. In Figure 4, the gateway IED is directly connected to the sub-controller, and its service capability is migrated from the master SDN controller. The sub-controller framework can be installed in the gateway IED of the smart grid system to achieve resource-constrained AMI network management.

4.2. Potential Benefits of the SDN-Enabled Smart Grid

- Simplified system management: The main purpose of SDN is to realize simplified network management that allows an extension of the scale and flexibility of a network. To reduce the management complexity of the network system, SDN isolates the control and data forwarding planes [15]. The control plane of the network is implemented using an SDN controller that constructs the network topology, and the data forwarding plane of the network is implemented using feedback information from the SDN devices on the network. The feedback information between the SDN controller and the device is exchanged using a secure control channel that is established using network tunneling [29]. Service providers can deploy application and network functionalities, such as the routing strategy, topology and QoS policy, using a secure control channel. In the SDN-enabled smart grid HAN, resource-constrained AMI devices do not need to decide their own routing path, QoS or transmission strategies. Instead, the centralized SDN controller architecture determines the optimal configuration of the network, reducing the resource usage of the AMI devices in terms of power consumption and the number of control messages required [30].

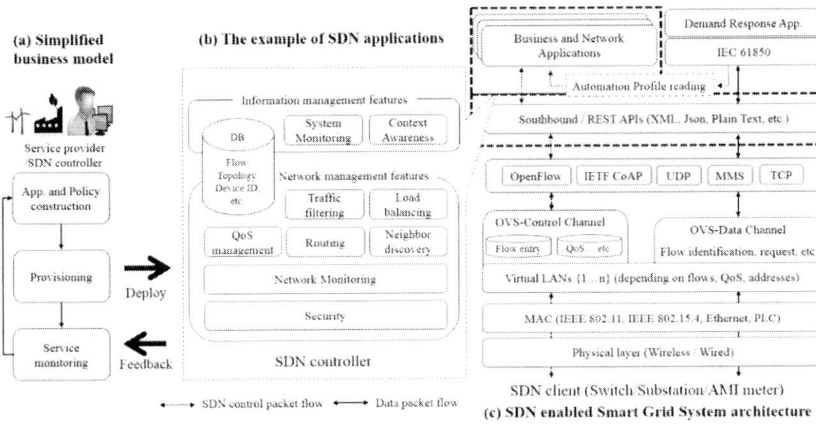

Figure 4. Abstraction of the SDN-enabled smart grid: business model process, categorized SDN application and system architecture. OVS, Open vSwitch; PLC, powerline communication. (**a**) Simplified business model. (**b**) The example of SDN applications. (**c**) SDN enabled Smart Grid System architecture.

The management software for the SDN controller can be categorized into two types, namely open-source-based and commercial systems. Nox [39] and Floodlight [12] are the most famous SDN control software frameworks that can handle the request and response messages that are sent between the application layer and network layer [39]. The SDN controller determines the network policy using an SDN application, and it maintains the flow information, route, forwarding strategy, topology and network status. Figure 4c shows an example of the SDN-enabled smart grid architecture. In this figure, SDN applications, such as network configuration, IEC 61850 configuration and system monitoring, are connected to the SDN controller using Southbound and REST APIs. Southbound is a connection application program interface (API) between the application and communication interface. When the AMI device receives a new flow, such as AMI metering data, it requests the forwarding policy from the SDN controller using the secure control channel. The forwarding policy of the requested flow is generated using the SDN application and is transmitted to the AMI meter. This process, which is carried out by the SDN-enabled smart grid, provides not only simplified network management, but also reduces the computational overhead of the AMI meter. Further, service providers simply add, remove or update network policies without the need for any hardware replacement or upgrade.

- Interoperability: SDN provides data and control message exchange between different types of communication infrastructure by using tunneling and Open vSwitch (OVS) and OpenFlow standards [29]. Figure 4c shows an example of the SDN system architecture. In this figure, the IEC 61850 utility protocol of the application and OVS is connected through the REST APIs of the OpenFlow interfaces. OpenFlow is the de facto SDN standard software that provides an interface between the control and forwarding layers of the SDN. OpenFlow allows direct access to and manipulation of the forwarding plane of network devices, such as switches, routers, whether they are physical or virtual (hypervisor based) [28]. OVS is an alternative to the bridge module that has been part of the kernel since the 2.4 series. OVS enables communication interface virtualization and port mapping for SDN router construction. All of the information received from each module can be monitored using SDN applications, and the service provider can access and control the system without the isolated stack regulations of the device.
- Situational awareness and flexibility: In SDN, the service provider can monitor the real-time network situation by using feedback information obtained from each SDN device. For example, in a traditional smart grid system, a service provider may not understand the need to upgrade the reception stability of the metering data without the need for additional network analyzers, such as firewalls, packet inspection routers and systems, when new standard protocols are utilized. However, in SDN, the network analysis system can be provisioned by employing an Open Systems Interconnection Reference Model Layer 2–3 network-monitoring SDN application [29]. Meanwhile, the service provider can simply identify network-wide network situations, such as congestion, collisions, bottlenecks and the flow status of the data packet, without additional hardware installation. In addition, network functionalities, such as routing, load balancing and the QoS management function, can be dynamically optimized and provisioned using network-wide situation information.
- Simplified service deployment: In the SDN-enabled smart grid, the application and network control plane can be removed from devices, such as routers, switches and AMI devices. Figure 4a–c shows an example of the SDN-enabled smart grid business model, SDN applications and system architecture. Generally, the SDN controller is directly managed by the service provider, and it is located in the HES. The service application is embedded within the application layer, which ensures hardware-independent system programming, and the new services and network policies can be dynamically deployed to the SDN

devices without the need for new hardware installation and configuration.

- Simplified business model: The SDN-enabled smart grid business model can be simplified to include policy development, service provisioning and service monitoring, as shown in Figure 4a. In this model, the system administrator develops an SDN application, such as topology construction, routing, QoS and traffic-filtering program. The implemented SDN application and policy need to be provisioned to the SDN controller, which is a part of the HES of the SDN-enabled smart grid system. In the application provisioning step, the service provider deploys an SDN application or policy to the SDN controller. The service monitoring is a key to developing a new application and profile. In this step, the service provider analyzes the network event history by using diagnostic SDN applications.

4.3. Candidate SDN Application for the Smart Grid

Figure 4b shows an example of the SDN applications. The SDN application is directly connected to the SDN stack using Northbound APIs [31]. Thus, the service providers do not need to consider hardware dependability, which allows interoperable data exchange between different network devices. The policy applies to the configuration profile of the SDN application, which enables the situational awareness of the smart grid infrastructure without the need for a replacement SDN application or hardware reconfiguration. The related studies about the information and network management SDN-enabled smart grid applications are categorized and shown in Table 1.

Table 1. Comparison between candidate SDN smart grid architectures.

Evaluated Features / Architecture	Context Awareness	System and Network Monitoring	Neighbor Discovery and Routing	Load Balancing	QoS	Security	SDN Controller Program	Network Interface(s)	Target Network
Molina, E. *et al.* [12]	√	√	-	√	-	√	Floodlight	Ethernet	WAN/NAN
Sydney, A. *et al.* [36]	-	√	-	√	√	-	Nox	Ethernet	WAN/NAN
Byun, J. *et al.* [33]	√	√	-	√	√	-	Custom	ZigBee/PLC	HAN
Andrew, *et al.* [13]	-	-	√	-	-	-	Nox	Ethernet	WAN
Cahn, A. *et al.* [14]	-	√	-	√		√	Custom	Ethernet	NAN
Qin, Z. *et al.* [38]	√	√	-	-	√	-	Qualnet, Sim	IEEE 802.11	HAN
Dorsch, N. *et al.* [34]	√	√	√	-	√	-	Custom	Ethernet	WAN/NAN
Dong, X. *et al.* [44]	-	√	-	-	-	√	ns-2, Sim	Ethernet	WAN/NAN

- Context awareness: Context awareness is an important feature for intelligent information management and provides situation-based interaction between a service provider and various IEDs. For example, the context awareness system can automatically identify and configure a new device's installation depending on its location and equipment type. In addition, this information can be utilized to anticipate an end-user's immediate needs, in the process offering more sophisticated, situational-aware, and usable functions. Magoutas, B. proposed a situational-aware demand-response algorithm for a smart meter that can be ported to the SDN controller [32]. In this algorithm, the system (controller) predicts the demand request patterns of the customer and automatically sends related information. Thus, the number of request messages and, thus, network congestion can be reduced. Byun, J. et al. proposed a smart energy distribution and management system (SEDMS). In this scheme, monitoring data from each household are reported to the centralized demand management system (DMS), which categorizes and optimizes the monitoring pattern for each household. The SEDMS application system is implemented using a software program without any specialized device, and its functions, such as the knowledge repository, context analyzer and monitoring-pattern generation module, can be dynamically updated or replaced by a new system without the need for any hardware replacement. These optimized traffic controls using the context-awareness function of the SDN can help to improve the network and service reliability of the resource-constrained smart grid infrastructure as an AMI system [36]. The experimental result of the SEMDS shows that the number of request messages and response delay time of the situational-awareness system are reduced compared to the traditional pre-defined profile-based demand response smart grid system [33].

- System and network monitoring: In traditional smart grid communication infrastructure, the service provider cannot monitor the network status without an additional packet inspection device. IEC 61850 focuses on device monitoring automation and not on network-wide monitoring. Thus, the service provider cannot determine the accruable reason for the instability of the communication infrastructure without using a network-monitoring device. Further, it may be difficult for the network system administrator to determine the actual problem leading to the loss of application packets, because the network stack cannot open the application packet itself. Thus, the system administrator can only identify the end-to-end perspective problem, even if the packet-loss problem occurred during packet transmission in the link layer (hop-by-hop). However, in the SDN-enabled smart grid, the service provider can not only monitor the device, but can also monitor the network status, such as the real-time throughput, dropped packets, topology and link-connection stability between devices. To realize application and network monitoring in SDN, the smart grid application characteristics and network parameters should be handled by an SDN controller. In 2014, Molina, E. et al. proposed a smart grid SDN management application for an IEC 61850-based smart grid system [12]. In this system, they presented a substation configuration description (SCD) information-mapping algorithm for SDN. In traditional smart grid infrastructure, the communication system cannot utilize these types of information for network optimization and monitoring. The SCD is an application profile of the IEC 61850 system and indicates various metering application profiles, such as monitoring intervals, attributes and types. However, in this scheme, the SCD profile is converted to the OpenFlow flow structure. The converted flow structure from the SCD is inserted into the OpenFlow entry table, and the SDN controller maintains the smart grid data flow using its monitoring application. Thus, the system administrator can simplify the access and management of both the application (service) and network using a single SDN controller. The experimental result of the IEC 61850 and OpenFlow achieves automated network monitoring as a real-time topology view, link bandwidth, traffic flow and address information of the substation network. Further, the authors can easily detect service denial attacks using the application and network monitoring SDN application without the need for additional network monitoring devices, such as firewalls [44].

- Neighbor discovery and routing: Optimal network configurations, such as routing and network topologies, are key to providing stable smart grid data communication links between HES and IEDs, but they cannot be dynamically configured by service providers in traditional networks. For example, the neighbor-discovery message interval cannot be dynamically configured in a traditional smart grid, even if the communication environment is very stable. Therefore, the service application may suffer from excessive control message overheads. In addition to the topology construction overhead, excessive routing-control message overheads may lead to a low QoS in resource-constrained AMI network environments [45]. SDN provides a flow-based forwarding strategy that employs combined neighbor discovery and routing SDN applications. Nils et al. proposed a multi-layered SDN4 smart grid architecture in 2014 [34]. In this architecture, the routing path between the central server and devices was constructed using a centralized SDN controller instead of the distributed routing protocol of the devices in the network. The SDN controller periodically monitors the network device, and constructs the network topology and route information between devices in the network. In addition, a SDN4 smart grid controller can maintain multiple routing paths between source and destination devices to provide DiffServ QoS routing between flows. If an AMI device, which has no route for incoming flow packets, requests the forwarding direction of the packet to the SDN controller, the SDN controller responds with the forwarding direction (next hop) to the requester. The experimental result of the end-to-end smart grid traffic delay shows that SDN4 smart grid routing outperforms open shortest path first (OSPF).

- Load balancing: SDN load balancing applications are flow information-based stateless control service applications that allow systems to bypass the limitations of traditional stateful hardware appliances, such as load balancers and firewalls [35]. In particular, load-balancing features must be considered for use in resource-constrained substation automation domains, such as wireless smart grid AMI networks. For example, if the emergency power-quality alert message from the AMI device is transmitted to the HES, end-to-end transmission needs to be guaranteed, but traditional network systems cannot detect specific flow priorities, because application profile information is not shared between application network layers. Therefore, the network cannot suspend low-priority flow transmission, even if the network capacity is overloaded. While traditional per packet-based admission control algorithms can be utilized to process network control message exchange, they need to be extended for application-flow identification. Sydney A. et al. proposed flow-based admission control for SDN [36]. In this algorithm, the SDN application creates two virtual network interfaces to isolate high-priority and best-effort flow transmission. The data rate of the developed virtual interfaces can be dynamically configured using queue-size management, which enables us to control the volume of the flow traffic in each network. This algorithm can be effectively utilized to protect the minimum bandwidth requirements of the high-priority data flow as emergency power quality traffic when a large volume of the best-effort data traffic, such as firmware and diagnostics flows, coexists on the network. The experimental result shows that the SDN controller can provide per flow-based service differentiation using a smart grid service profile.

- QoS management: The smart grid communication infrastructure must simultaneously support a wide range of QoS requirements for numerous smart grid and legacy applications. To satisfy the delay requirement of the smart grid traffic, the SDN controller uses both the context-awareness and load-balancing SDN applications. In the traditional network infrastructure, cross-layer-based QoS algorithms are widely utilized, but their use increases the network-system complexity [37]. However, in SDN, cross-layer information is easily shared using OpenFlow and OVS, which are operated without any direct information exchange between different network layers [29]. Qin, Z. et al. proposed a heuristic flow scheduling-based QoS algorithm for smart grid devices [38]. In this algorithm, the authors assumed that SDN can detect the flow information and real-time flow status using an SDN flow monitoring application. The SDN controller assigns flow priority depending on a calculus-based queue model that considers the flow-traffic arrival rate, delay requirement of the flow and the available network bandwidth of the device. Each application flow of the AMI device can be assigned a different virtual interface depending on the smart grid service profile. The interface configuration can also be dynamically changed using the calculated flow priority of the SDN controller to optimize the network capacity and service QoS. The performance result shows that SDN APP can provide not only the DiffServ QoS of the smart grid flows, but also the dynamic QoS configuration of the flows in the network.

- Security: The smart grid system autonomously collects massive amounts of data and transfers them to utility companies, consumers and service providers [35]. These data include the private information of customers, such as activities, devices being used and times at which the home is vacant. Abnormal user access and control packets of the SDN are protected by secure socket layer (SSL) and digital mark-based packet-inspection algorithms. The traditional packet-inspection system that is used as a firewall may be degraded because of the large network scale of the smart grid. In addition, the network installation cost will increase when the number of firewalls in the network is increased. However, SDN can provide packet inspection without specialized hardware, which allows dynamic protection rule updates by a system administrator. The ForNox is a role-based access control system for SDN [39]. In this scheme, the SDN controller inserts the validated digital sign to all control packets, and it is encapsulated by using a pre-shared key. Further, the SDN control channels are protected by the SSL protocol [29]. Rowan et al. presented an SDN security analysis model for information disclosure, denial of service and tampering attack detection [40]. In this algorithm, the SDN controller randomly chooses the sample packets and data flow table from the network devices. The collected sample data are analyzed using Microsoft STRIDE [40] and attack tree modeling methods, and experimental results show that the

SDN effectively isolates and detects the abnormal flows without the need for an additional network monitoring system.

5. FURTHER RESEARCH OPPORTUNITIES OF THE SDN-ENABLED SMART GRID

SDN is an emerging and promising future networking solution, but it needs to be optimized for use in a smart grid environment. Traditional SDN is designed for rich resource systems that are generally considered for backhaul IP networks. From a network design perspective, SDN-enabled smart grid systems require an additional network processing server, which must consider the end-to-end route between the SDN controller and each smart grid device in the region. Therefore, the communication overhead between the SDN controller and devices needs to be optimized in order to reduce network congestion, bottlenecks and computational delay problems. Challenging issues pertaining to the SDN-enabled smart grid system are as below:

- Reducing SDN control message overhead: In spite of the different benefits, such as simplified network and system management, SDN may not be directly utilized in resource-constrained AMI networks because of the control message and end-to-end session-management overhead. Generally, the SDN controller and all clients periodically exchange control messages for service provisioning and monitoring via a secure path. These control messages use secure control paths that are logically isolated from the data path using SSL or virtual private networks (VPNs). However, the MAC and physical perspective of the communication interface control and data packet use the same physical interface. Further, SSL-based data communication requires end-to-end session management control packets that are periodically exchanged between the SDN controller and all devices in the network. From a security perspective, control and data channel isolation are acceptable methods, but the end-to-end transmission reliability of the data packet can suffer from network congestion due to the SDN control traffic. In addition, when the control message between the controller and device is not stably delivered, metering traffic cannot be transmitted, even if the network capacity is sufficient. In particular, resource-constrained smart grid networks, such as wireless HANs, need to consider low-overhead message exchange protocols, such as Internet Engineering Task Force (IETF)-Constrained Application Protocol (CoAP) , to reduce congestion and realize a reliable SDN-enabled smart grid [41,45].

- Route construction between SDN controller and clients: SDN enables flow-based data forwarding that can replace the traditional distributed routing protocols. This centralized routing is a key feature of SDN, but during the network initialization stage, each client needs to find a default route path to the SDN controller [42]. To solve this problem, a distributed routing protocol, such as OSPF, is considered for simultaneous use with SDN in wired SDN, but it is not suitable for use in resource-constrained wireless and PLC-based HANs. To solve this problem, the SDN-enabled smart grid must use the default routing function, which provides a dynamic configuration to legacy routing protocols, or discover the routing path by using an SDN routing application. In HANs, the routes between the SDN controller and a device are constructed using a link discovery procedure. Generally, the SDN controller uses a link discovery application that enables it to find the adjacent one-hop neighbor SDN controller of the device. Meanwhile, the device that is connected to the SDN controller can respond to the link discovery request with SDN connection flags. This connection information can be sequentially flooded to the entire network, and the default routes between the SDN controller and the devices are established without additional routing protocols.
- SDN controller and function distribution: In a large-scale smart grid network, the flow information request messages from all clients are concentrated at an SDN controller. There is the potential problem of network congestion, queue overflow, large processing delay and service instability of the SDN. To solve this problem, multiple SDN controllers need to be distributed to the network and HES. In the case of multiple SDN controllers, all of them need to share the common management policy of the network. For example, the elastic SDN controller distribution (ElastiCon) algorithm can be utilized for a large-scale smart grid network [43]. In this algorithm, the SDN controllers are classified as a master and slaves. The master SDN controller can be located in the HES, and slaves are inserted at the substations. The master SDN controller assigns and distributes the slave SDN controllers to the network depending on the link capacity of the switch or router. Further, the SDN network functions (application) are distributed to the multiple slaves. During the migration process between the master and sub-controllers, at least one controller is activated to avoid a service suspension problem. The experimental results show that the network congestion, end-to-end control message delay, flow request-response time problem can be solved by using an elastic controller distribution system.

- Interoperable design between IEC 61850 and SDN protocols: In traditional layered networking architecture, communication strategies, such as routing policy, QoS policy and MAC parameters, may not dynamically adapt to applications and services. However, an SDN-enabled smart grid network can be optimized for specific smart grid services by using IEC 61850 standard profiles. The IEC 61850 protocol is the de facto application standard of the smart grid [12]. The information model of the smart grid substation is defined in IEC 61850-7, and the configuration language is also defined in IEC 61850-6. The IEC 61850 application defines the power-monitoring interval, transmission type (IEC 61850 Manufacturing Message Specification or multicast) and service priorities. The IEC 61850 uses the XML (or JSON) data format, which can be decoded using SDN applications because the SDN API basically uses the XML and JSON formats to exchange control messages between the SDN controller and clients. Thus, the service profile of the IEC 61850 can be simply identified and utilized for the network configuration. By using an interoperable management architecture between IEC 61850 and SDN, the service capability of the SDN-enabled smart grid can be extended from the network to smart grid application management.
- Interoperable controller system: Interoperability among different SDN architectures needs to be considered for SDN-enabled smart grid systems [45]. This is because SDN is implemented by a vendor-neutral strategy, depending on the capability and business goal of the service domain, without the need for any international agreement. In particular, northbound API and SDN controller systems need to be standardized to realize simplified communication and the development of smart grid SDN applications.

6. CONCLUSIONS

The complexity and heterogeneity of smart grid services are rapidly increasing because of the increased usage of IEDs, automation and service integration between utility domains. However, traditional networks that focus on distributed packet-forwarding strategies may not dynamically adapt to different business models, because the context and situation status of the application cannot be directly utilized for network configuration. Instead of the legacy networking system, in recent years, the SDN paradigm has been considered for the smart grid network infrastructure. However, traditional SDN architectures need to be optimized for the smart grid environment and this process is still at the initial stages.

In this paper we investigated the current information subsystem, communication infrastructure of the smart grid and an emerging communication infrastructure called the SDN-enabled smart grid. In particular, we described the SDN-enabled smart grid architecture in detail, as well as the abstracted business model process of the service provider, a candidate SDN-enabled smart grid application and system architecture. Finally, we presented a comparison of current SDN-enabled smart grid systems and categorized the examined network services and applications that have been developed based on the SDN paradigm. To the best of our knowledge, this is the first such comparison of studies into SDN-enabled smart grid communication infrastructures. Finally, we identified challenging issues affecting the current SDN-enabled smart grid and proposed possible solutions.

ACKNOWLEDGMENTS

This publication was made possible by the National Priorities Research Programme (NPRP) award (NPRP 6-244-2-103) from the Qatar National Research Fund (a member of The Qatar Foundation). The statements made herein are solely the responsibility of the author(s).

AUTHOR CONTRIBUTIONS

Jaebeom Kim and Fethi Filali conceived of the principle reported in this work for the current smart grid architecture and its trends. Young-Bae Ko categorized the current ICT subsystems of the smart grid and its sub-technologies from communication infrastructure and services. Jaebeom Kim and Fethi Filali jointly developed an abstract SDN-enabled smart grid architecture model and its related work. Jaebeom Kim wrote the paper.

CONFLICTS OF INTEREST

The authors declare no conflict of interest.

REFERENCES

1. Locke, G.; Gallagher, P.D. Nist Framework and Roadmap r mart Grid Interoperability Standards, Release 1.0; National Institu o Standards and Technology: Gaithersburg, MD, USA, January 2010.
2. Güngör, V.C.; Sahin, D.; Kocak, T.; Ergüt, S.; Buccel C.; Cecati, C.; Hancke, G.P. Smart Grid technologies: Communication hnologies and standards. IEEE Trans. Ind. Inform. **2011**, 7, 529–539.
3. Liserre, M.; Sauter, T.; Hung, J.Y. Future energy sy ms: Integrating renewable energy sources into the smart power grid rough industrial electronics. IEEE Ind. Electron. Mag. **2010**, 4, 18–37.
4. Hauser, C.H.; Bakken, D.E.; Bose, A. A failure to mmunicate: Next generation communication requirements, chnologies, and architecture for the electric power grid. IEEE Pow Energy Mag. **2005**, 3, 47–55.
5. Fang, X.; Misra, S.; Xue, G.; Yang, D. Smart Grid—e new and improved power grid: A survey. IEEE Commun. Surv. Tutor **012**, 14, 944–980.
6. Kim, J.; Kim, D.; Lim, K.-W.; Ko, Y.-B.; Lee, S.-Y. Iproving the reliability of IEEE 802.11 s based wireless mesh network or Smart Grid systems. J. Commun. Netw. **2012**, 14, 629–639.
7. Niyato, D.; Xiao, L.; Wang, P. Machine-to-ma ine communications for home energy management system in Smart rid. IEEE Commun. Mag. **2011**, 49, 53–59.
8. Bauer, M.; Plappert, W.; Wang, C.; Dstert, K. Packet-oriented communication protocols for Smart Grid srvices over low-speed PLC. In Proceedings of the IEEE Internationa Symposium on Power Line Communications and Its Applications, IS LC 2009, Dresden, Germany, 29 March–1 April 2009; pp. 89–94.
9. Kuzlu, M.; Pipattanasomporn, M.; Rahnan, S. Communication network requirements for major Smart Grid applications in han, nan and wan. Comput. Netw. **2014**, 67, 74–88.
10. Yi, P.; Iwayemi, A.; Zhou, C. Frequency agility in a zigbee network for Smart Grid application. In Proceedings of the Innovative Smart Grid Technologies (ISGT), Gaithersburg, MD, USA, 19–21 January 2010; pp. 1–6.
11. Bumiller, G.; Sauter, T.; Pratl, G.; Treytl, A. Secure and reliable wide-area power-line communication for soft-real-time applications within

REMPLI. In Proceedings of the 2005 International Symposium on Power Line Communications and Its Applications, Vancouver, Canada, 6–8 April 2005; pp. 57–60.

12. Molina, E.; Jacob, E.; Matias, J.; Moreira, N.; Astarloa, A. Using software defined networking to manage and control IEC 61850-based systems. Comput. Electr. Eng. **2015**, 43, 142–154.

13. Goodney, A.; Kumar, S.; Ravi, A.; Cho, Y.H. Efficient PMU networking with software defined networks. In Proceedings of the 2013 IEEE International Conference on Smart Grid Communications (Smart GridComm), Vancouver, BC, Canada, 21–24 October 2013; pp. 378–383.

14. Cahn, A.; Hoyos, J.; Hulse, M.; Keller, E. Software-defined energy communication networks: From substation automation to future Smart Grids. In Proceedings of the 2013 IEEE International Conference on Smart Grid Communications (Smart GridComm), Vancouver, BC, Canada, 21–24 October 2013; pp. 558–563.

15. Zhang, J.; Seet, B.-C.; Lie, T.-T.; Foh, C.H. Opportunities for software-defined networking in Smart Grid. In Proceedings of the 2013 9th International Conference on Information, Communications and Signal Processing (ICICS), Tainan, Tanwan, 10–13 December 2013; pp. 1–5.

16. Parniewicz, D.; Doriguzzi Corin, R.; Ogrodowczyk, L.; Rashidi Fard, M.; Matias, J.; Gerola, M.; Fuentes, V.; Toseef, U.; Zaalouk, A.; Belter, B. Design and implementation of an openflow hardware abstraction layer. In Proceedings of the 2014 ACM SIGCOMM workshop on Distributed cloud computing, Chicago, IL, USA, 17–22 August 2014; pp. 71–76.

17. Pullins, S. West Virginia Smart Grid Implementation Plan; Technical Report DOE/NETL-2009/1386. West Virginia Division of Energy, National Energy Technology Laboratory, US Department of Energy: Morgantown, WV, USA, November 2009.

18. Logenthiran, T.; Srinivasan, D.; Shun, T.Z. Demand side management in Smart Grid using heuristic optimization. IEEE Trans. Smart Grid **2012**, 3, 1244–1252.

19. Pendarakis, D.; Shrivastava, N.; Liu, Z.; Ambrosio, R. Information aggregation and optimized actuation in sensor networks: Enabling smart electrical grids. In Proceedings of the 26th IEEE International Conference on Computer Communications, INFOCOM 2007, Anchorage, AK, USA, 6–12 May 2007; pp. 2386–2390.

20. Luan, W.; Sharp, D.; Lancashire, S. Smart Grid communication network capacity planning for power utilities. In Proceedings of the 2010 IEEE

PES Transmission and Distribution Conference and Exposition, New Orleans, LA, USA, 19–22 April 2010; pp. 1–4.

21. Vytelingum, P.; Voice, T.D.; Ramchurn, S.D.; Rogers, A.; Jennings, N.R. Agent-based micro-storage management for the Smart Grid. In Proceedings of the 9th International Conference on Autonomous Agents and Multiagent Systems, Toronto, ON, Canada, 9–14 May 2010; Volume 1, pp. 39–46.

22. Momoh, J. Smart Grid design for efficient and flexible power networks operation and control. In Proceedings of the PSCE 2009 IEEE/PES Power Systems Conference and Exposition, Seattle, WA, USA, 15–18 March 2009; pp. 1–8.

23. Moreno-Munoz, A.; de La Rosa, J.G. Integrating power quality to automated meter reading. IEEE Ind. Electron. Mag. **2008**, 2, 10–18.

24. Meehan, B. Enterprise Gis and the Smart Electric Grid; Electric Power White Paper, ESRI; 2008.

25. Jeremy, P. Optimizing the Performance of Smart Grid Technologies; Business Insights White Paper, Pitney Bowse; 2010.

26. Bari, A.; Jiang, J.; Saad, W.; Jaekel, A. Challenges in the smart grid applications: An overview. International Journal of Distributed Sensor Networks **2014**, 2014, 1–11.

27. Giordano, V.; Fulli, G. A business case for Smart Grid technologies: A systemic perspective. Energy Policy **2012**, 40, 252–259.

28. Hashmi, M.; Hänninen, S.; Mäki, K. Survey of Smart Grid concepts, architectures, and technological demonstrations worldwide. In Proceedings of the 2011 IEEE PES Conference on Innovative Smart Grid Technologies (ISGT Latin America), Medellin, Colombia, 19–21 October 2011; pp. 1–7.

29. Shirazipour, M.; John, W.; Kempf, J.; Green, H.; Tatipamula, M. Realizing packet-optical integration with sdn and openflow 1.1 extensions. In Proceedings of the 2012 IEEE International Conference on Communications (ICC), Ottawa, ON, Cananda, 10–15 June 2012; pp. 6633–6637.

30. Jarschel, M.; Wamser, F.; Hohn, T.; Zinner, T.; Tran-Gia, P. Sdn-based application-aware networking on the example of youtube video streaming. In Proceedings of the 2013 Second European Workshop on Software Defined Networks (EWSDN), Berlin, Germany, 10–11 October 2013; pp. 87–92.

31. Salsano, S.; Blefari-Melazzi, N.; Detti, A.; Morabito, G.; Veltri, L. Information centric networking over sdn and openflow: Architectural aspects and experiments on the ofelia testbed. Comput. Netw. **2013**, 57, 3207–3221.

32. Magoutas, B.; Apostolou, D.; Mentzas, G. Situation-aware demand response in the Smart Grid. In Proceedings of the 2011 16th International Conference on Intelligent System Application to Power Systems (ISAP), Hersonissos, Greece, 25–28 September 2011; pp. 1–6.

33. Byun, J.; Hong, I.; Kang, B.; Park, S. A smart energy distribution and management system for renewable energy distribution and context-aware services based on user patterns and load forecasting. IEEE Trans. Consum. Electron. **2011**, 57, 436–444.

34. Dorsch, N.; Kurtz, F.; Georg, H.; Hagerling, C.; Wietfeld, C. Software-defined networking for Smart Grid communications: Applications, challenges and advantages. In Proceedings of the 2014 IEEE International Conference on Smart Grid Communications (Smart GridComm), Venice, Italy, 3–6 November 2014; pp. 422–427.

35. Wang, X.; Yi, P. Security framework for wireless communications in smart distribution grid. IEEE Trans. Smart Grid **2011**, 2, 809–818.

36. Sydney, A.; Ochs, D.S.; Scoglio, C.; Gruenbacher, D.; Miller, R. Using geni for experimental evaluation of software defined networking in Smart Grids. Comput. Netw. **2014**, 63, 5–16.

37. Cuomo, F.; Abbagnale, A.; Cipollone, E. Cross-layer network formation for energy-efficient IEEE 802.15.4/zigbee wireless sensor networks. Ad Hoc Netw. **2013**, 11, 672–686.

38. Qin, Z.; Denker, G.; Giannelli, C.; Bellavista, P.; Venkatasubramanian, N. A software defined networking architecture for the internet-of-things. In Proceedings of the 2014 IEEE Network Operations and Management Symposium (NOMS), Krakow, Poland, 5–9 May 2014; pp. 1–9.

39. Pfeiffenberger, T., Du, J.L. Evaluation of software-defined networking for power systems. In Proceedings of the 2014 IEEE International Conference on Intelligent Energy and Power Systems (IEPS), Kyiv, Ukraine, 2–6 June 2014; pp. 181–185.

40. Kloti, R.; Kotronis, V.; Smith, P. Openflow: A security analysis. In Proceedings of the 2013 21st IEEE International Conference on Network Protocols (ICNP), Goettingen, Germany, 7–10 October 2013; pp. 1–6.

41. Castellani, A.P.; Gheda, M.; Bui, N.; Rossi, M.; Zorzi, M. Web services for the internet of things through CoAP and EXI. In Proceedings of the 2011 IEEE International Conference on Communications Workshops (ICC), Kyoto, Japan, 5–9 June 2011; pp. 1–6.

42. Das, S.; Parulkar, G.; McKeown, N. Why Openflow/SDN can succeed where GMPLS failed. In Proceedings of the 2012 38th European Conference and Exhibition on Optical Communication, Amsterdam, The Netherlands, 16–20 September 2012; pp. 1–3.

43. Dixit, A.; Hao, F.; Mukherjee, S.; Lakshman, T.; Kompella, R. Towards an Elastic Distributed SDN Controller. In Proceedings of the second ACM SIGCOMM workshop on Hot topics in software defined networking, Hong Kong, China, 12–16 August 2013; pp. 7–12.

44. Dong, X.; Lin, H.; Tan, R.; Iyer, R.K.; Kalbarczyk, Z. Software-defined networking for Smart Grid resilience: Opportunities and challenges. In Proceedings of the 1st ACM Workshop on Cyber-Physical System Security, Singapore, 14–17 April 2015; ACM: New York, NY, USA; pp. 61–68.

45. Jaebeom, K.; Y-B, K.; Fethi, F. A Lightweight CoAP-based Software Defined Networking for Resource Constrained AMI Devices. In Proceedings of the IEEE International Conference on Smart Grid Communications (Smart GridComm), Miami, FL, USA, 2–5 November 2015; pp. 1–6.

46. Fujitus. Software-Defined Networking for the Utilities and Energy Sector; Fujitus White Paper, Fusitus; 2014.

11

CHAPTER

INVESTIGATION OF THE PERFORMANCE OF CIS PHOTOVOLTAIC MODULES UNDER DIFFERENT ENVIRONMENTAL CONDITIONS

Kandil M. Kandil, Majida S. Altouq, Asma M. Al-asaad, Latifa M. Alshamari, Ibrahim M. Kadad, Adel A. Ghoneim*

Applied Sciences Department, College of Technological Studies, Shuwaikh, Kuwait.

ABSTRACT

This work investigates the effects of temperature and radiation intensity on the parameters of a copper indium diselenide (CIS) photovoltaic module. The module performance parameters are determined from calculated module parameters. An outdoor experimental setup is installed to carryout a series of I-V curve measurements under different irradiance and temperature conditions for the module. A numerical model which considers the effect of series and shunt resistances is developed to evaluate the different parameters of PV modules. Orthogonal distance regression (ODR) algorithm is adapted for fitting I-V measurements and extracting module parameters from I-V measurements. The values of module parameters, series resistance R_s, shunt resistance R_{sh}, diode ideality factor n and reverse saturation current I_o determined from I-V measurements at different irradiation intensity and temperature range are in good agreement with the corresponding parameters obtained from the developed numerical model. The module parameters extracted from I-V measurements are employed to calculate the module performance parameters, i.e. open circuit voltage V_{oc}, fill factor FF and module efficiency h at different irradiation intensity and temperature range. Present results indicate that the largest drop in open circuit voltage V_{oc} due to about 20°C increase in temperature is approximately 8.8% which is not compensated for by the relatively small increase in short circuit current, (2.9% in Isc), resulting in a reduction in maximum power of about 6.3%. Results let us conclude that the shunt resistance R_{Sh} increases with radiation at low radiation values (<400 W/m^2). As radiation increases at high radiation values (>400 W/m^2), R_{Sh} begins to decease sharply and dramatically. Also, as the light intensity incident on the solar module increases, the series resistance and the output voltage decrease. When the irradiance intensity increases, the series resistance decreases but with a very low rate at the two studied temperatures ranges. The low rate decrease of R_s is found to have little effect on module performance in comparison with the significant change of other module parameters. The ideality factor n and saturation current I_o decrease first sharply in the low range of radiation intensity (<400 W/m^2) and this decrease becomes smaller for irradiance values greater than 400 W/m^2. The previous observations and conclusions regarding the module parameters R_{Sh}, R_S, n and I_o obtained at 20°C observed again at 40°C but there is a great difference between the peak values of R_{Sh} at both temperature levels. Present results also show that module efficiency decreases with increasing irradiance intensity due to the combined effect of both V_{oc} and FF.

Keywords: PV Module, Module Parameters, Series Resistance, Shunt Resistance, Ideality Factor

1. INTRODUCTION

The main objective of photovoltaic research industry is to develop high efficiency low cost photovoltaic cells and modules. Photovoltaic technology clearly offers tremendous environmental benefits, requiring no fuel and producing no emissions or other waste beyond that inherent in the manufacturing process. Moreover, photovoltaic have proven to be economical for a wide range of applications that have traditionally relied on diesel generators. The advantages that photovoltaic have over competing power options are: they have no moving parts and produce power silently. They are non-polluting with no detectable emissions or odors. They can be standalone systems that reliably operate unattended for long periods. They require no connection to an existing power source or fuel supply. They consume no fossil fuels-their fuel is abundant and free. Unfortunately solar cells are still far too expensive to produce a significant fraction of the worlds energy needs. The use of photovoltaic cells has increased in the last few decades as their manufacturing cost has decreased and as people have become more concerned about energy use.

A thin-film solar cell is a solar cell that is made by depositing one or more thin layers of photovoltaic material on a substrate. The thickness range of such a layer is wide and varies from a few nanometers to tens of micrometers. Many different photovoltaic materials are deposited with various deposition methods on a variety of substrates. Thin-film solar cells are usually categorized according to the photovoltaic material used: amorphous silicon (a-Si), cadmium telluride (CdTe), copper indium gallium selenide (CIS or CIGS) and dye-sensitized solar cell (DSC) and other organic solar cells. Thin film solar cells have achieved efficiencies of about 15 - 20%. The highest efficiencies recorded recently for all thin film photovoltaic technology on laboratory scale or module scale are for chalcopyrite solar cell based on Cu $(InGa)Se_2$ absorbers. So, obtaining the outdoor performance of the base material Cu In Se_2 (CIS) or the more efficient one Cu $(InGa)Se_2$ (CIGS) under different solar irradiance and device temperature and or spectral irradiance is vital for selecting the proper solar module for certain location and specific application.

Series and shunt resistances in solar cells are parasitic parameters, which significantly affect the illuminated I-V characteristics and cell efficiency. Series resistance, R_s, is mainly the sum of contact resistance on the front and back surfaces. So, the series resistance should be kept to a minimum for efficient conversion of solar energy. The shunt resistance, R_{sh}, represents a parallel high conductivity path across the p-n junction or the cell edges and decreases the efficiency of the cells by increasing the leakage current that lowers the maximum output power (P_m), the open voltage, V_{oc} and the fill factor, FF. Several methods are available in the literature for extracting the series and shunt resistances of a solar cell [1-12] as well as related device parameters. Some of the methods involve measurement of illuminated I–V characteristics, some use dark conditions or utilize dark and illumination measurements. Other methods apply curve fitting method to some assumed functional relationship or employ integration procedures based on the computation of the area under the I-V curves or use linear regression. It is also possible to predict PV module performance under illumination through a several numerical or algebraic methods [13-18].

Most of the above mentioned methods are based on single exponential model of solar cell and assume that the shunt resistance, R_{sh}, is infinite and presume that R_s is independent of the intensity of illumination which is not a correct assumptions especially for solar cells that have low R_{sh} as amorphous silicon and CIS modules.

For this reason, an improved five parameter model which considers the shunt resistance is developed to carry out the present work. The model can handle the different cell technologies namely amorphous crystalline and copper indium diselenide (CIS) thin-film modules. The five parameter model uses the data provided by the manufacturers and empirical correlation equations to predict the solar cell parameters at different operating conditions. Measurements of I-V characteristics are carried out in the photovoltaic facility installed at the roof of the main building in the College of Technological Studies, Kuwait. These measurements are carried out at different illumination intensity and temperature. Non linear curve fitting of experimental I-V data proposed by Burgers et al. [19] is adapted to extract different solar modules parameters from I-V measurements. A comparison between solar modules parameters predicted from the developed numerical model and the corresponding parameters obtained from I-V measurements to validate the accuracy of the developed numerical model proposed for this study. The module parameters extracted from I-V measurements are employed to calculate the module performance parameters, i.e. short circuit current (I_{sc}) open circuit voltage (V_{oc}), fill factor (FF) and efficiency (h) at different irradiation intensity and temperature range.

2. THEORETICAL ANALYSIS

A PV cell may consist of a single crystal, a number of smaller crystals (polycrystalline), or it may lack crystal structure all together (amorphous). Other semiconducting materials such as thin film chalcopyrite are employed in PV cells for extraterrestrial or sun concentrator applications, but this description will concern a silicon PV cell. The light intensity affects primarily the amount of current produced, making it proportional, while the cell temperature controls the voltage produced. As the cell temperature increases, the current produced remains the same but the voltage is reduced, reducing the output power. All of these factors need to been taken into consideration to accurately predict the energy production.

To choose an appropriate model for detailed simulation, several factors need to be considered. The most important one is the accuracy which can be obtained. A PV cell is not a fixed voltage source, so the model should be able to predict the current and voltage over the entire operating voltage range. Another factor is whether the data required by the model is provided by the manufacturers or not. Another important consideration is whether the data needed to use an I-V model are readily available to the system designer. Models with a large number of parameters, although possibly more accurate, either requires access to generally unpublished and proprietary manufacturer's data, or else require prototype test data. The most detailed, consistent information commonly available is provided in brochures from commercial module manufacturers.

Standard test conditions (STC) are usually specified so that a performance comparison can be made between different PV units (cell, modules). The conditions for standard test conditions are specified as follows: The reference irradiance G_{Ref} with a typical value of 1000 W/m^2, the reference cell temperature $T_{C,Ref}$ with a typical value of 25°C and a specified light spectral distribution with an air mass, AM = 1.5. There are three classic parameters that are very important on the PV characteristics namely I_{sc}, V_{oc} and the maximum power point (I_{mp}; V_{mp}). These points are usually given as part of a manufacturer's data sheet for a PV module. Another important parameter of the PV performance is called the Fill Factor (FF). It is a term that describes how the curve fills the rectangle that is defined by (V_{oc}) and (I_{sc}). It gives an indication of the quality of a cell's semiconductor junction and measures of how well a solar cell is able to collect the carriers generated by light. It is defined as:

$$FF = \frac{V_{mp} I_{mp}}{V_{oc} I_{sc}} \tag{1}$$

After a simple manipulation, one gets the equation:

$$V_{\text{oc}} I_{\text{sc}}\ FF = V_{\text{mp}} I_{\text{mp}} = P_{\text{max}} \tag{2}$$

It can be easily observed that FF is always less than unity and differs from material to material. The closer the value of the fill factor is to unity, the better the operation of the PV cell. The PV efficiency (h) is defined as the ratio of the maximum output power P_{mp} to the solar power received by the cell surface area (A):

$$\eta = \frac{FFI_{sc}V_{oc}}{GA} \quad (3)$$

A five parameter photovoltaic model is adapted in this study to determine the module parameters. In the present work, the five parameter model is used to simulate the characteristic of thin film solar cells at different weather conditions. This model adds R_{sh} to the four-parameter model making it applicable to both crystalline and thin film PV solar cells.

The mathematical model describing the performance of the solar modules should accurately predict how solar module output would vary with ambient temperature and radiation conditions. Since the solar cell is a nonlinear power source, so the output current and voltage depend on the radiation level and temperature. There are different photovoltaic simulation programs developed for modeling the performance of the solar cell. In these models, the solar cell is treated as an equivalent circuit. Such models require a program to simultaneously determine the current-voltage (I-V) characteristics as well as the output of a PV module. The four parameters model is a widely used model for determining the characteristic curves of solar cells. The four-parameters appearing in the IV equation are the light current (I_L), the series resistance (R_S), and two theoretical diode characteristics. These parameters are not measurable quantities and are not generally included in manufactures catalog data. As a result, they must be determined from systems of I-V equations at various operating points; these points are taken from catalog data. The detailed analysis of the four-parameter PV model was introduced by Townsend [13] and incorporated into TRNSYS [14].

The four parameter model assumes that the slope of the I-V curve is flat at the short-circuit condition. This means that this model neglects R_{sh} because it is very large compared with R_s for the crystalline silicon (C-Si) cell. However, this assumption is not generally valid for amorphous photovoltaic. So, the four parameters model accurately predicts the performance of single crystal and polycrystalline PV arrays. In the other hand, the four parameters model can not accurately predict the performance of thin film PV modules. The main deviation is around the maximum power point because a-Si solar cell does not have a good fill factor as the C-Si cell does. In order to explain the low fill factors measured, the shunt resistance, which are lower than that of C-Si cell should be taken into consideration. Thee five parameters model takes R_s into account. Adding this parameter makes the five-parameter model applicable to both crystalline and amorphous PV cells. In the present work, a five parameter model is developed to simulate the characteristics of CIS module at different weather conditions.

The electrical behavior of the five parameter model is usually described by an equivalent circuit. The IV equation for five parameter model is generally expressed by:

$$I = I_L - I_o \times \left\{ \exp\left(\frac{q}{nkT_C}(V + IR_S) \right) - 1 \right\} - \frac{V + IR_S}{R_{sh}} \tag{4}$$

where I_L is the light current, I_o is the diode reverse saturation current, junction of the cell material. The diode current, I_D is the current flowing internally across the cell's semiconductor junction. All the internal dissipative electrical losses are lumped together into the series resistance, R_S. R_{sh} is the shunt resistance and as it approaches infinity, this arrangement becomes identical to the fourparameter model. Thus, all the equations employed in the five-parameter model reduce to those in the simpler four parameter scheme for very large values of R_{sh}. Equation (4) also includes ideality factor (n), electron charge (q), Boltzmann constant (k) and cell temperature (T_C). The five parameters from which the five parameter model obtains its name are: n, I_L, I_o , R_s and R_{sh}. Amorphous and CIS photovoltaic cells exhibit a pronounced slope at the short circuit point, therefore a finite shunt resistance exists. A numerical model is adapted to solve Equation (4) to obtain the five parameters at different illumination intensity and operating temperatures. The magnitude of the IV slope at short-circuit is related inversely to the shunt resistance. As R_{sh} approaches infinity, this arrangement becomes identical to the four-parameter model. Thus, all the equations employed in the five-parameter model reduce to those in the simpler four parameter scheme for very large values of R_{sh}.

An additional equation is needed to numerically solve IV equation. The catalog value that is employed to get the additional equation is the temperature coefficient of open-circuit voltage (m_{VOC}). Taking the derivative of Equation (4) and using an equation for the temperature dependence of I_o yields:

$$\mu_{Voc} = \frac{\delta V_{oc}}{\delta T_C} = \frac{\mu_{Isc} - \frac{I_{o,ref}}{T_C}\left(3 + \frac{q\varepsilon_g}{nkT}\right)\exp\left(\frac{q}{nkT_{C,ref}}\right)}{\frac{q}{nkT_{C,ref}} \cdot I_{o,ref}\left[\frac{q}{nkT_C} V_{oc,ref}\right] + \frac{1}{R_{sh}}} \tag{5}$$

The diode reverse saturation current (I_o) depends on temperature, as well as on the material band gap energy (ε_g) and the number of cells in series (Ns). So, when the value of saturation current increases, the I-V curve shifts toward the left decreasing the value of the open circuit voltage and the voltage at the maximum power point.

The light current (I_L) parameter was observed to depend on the effective solar irradiance, the cell temperature (T_c), the short circuit current temperature coefficient (m_{Isc}), and the air mass modifier. The equation for I_L is:

$$I_L = \frac{G}{G_{Ref}}\left[I_{L,Ref} + \mu_{ISC}\left(T_C - T_{C,Ref}\right)\right] \quad (6)$$

Subscript (Ref) refers to values at reference conditions.

2.1. Determining the Shunt Resistance

Adding R_{sh} introduces another unknown value into the model. Like I_L, I_o, n, and R_s, R_{sh} is not an easily measured physical quantity. Thus, it must be determined from information available from the manufacturer. The primary effect of R_{sh} on the IV characteristics is to change the short circuit slope. Thus, it is logical to use the measured value of this slope to back out a value for R_{sh}. The short-circuit slope is not a value which is generally included in the list of catalog specifications. However, it is possible to approximate this value if the IV curve is available. In order to determine the relationship between the short-circuit slope and the shunt resistance it is first necessary to differentiate the IV expression presented in Equation (4) with respect to voltage. Performing the differentiation, setting V to 0 and I to $I_{sc,ref}$ yields:

$$m = -\frac{q}{nkT_C} I_o\left(1 + R_S m\right)\exp\left(\frac{q}{nkT_C} I_{sc} R_s\right) - \frac{\left(1+m\right)R_s}{R_{sh}} \quad (7)$$

where the variable m is equal to the short circuit slope $\left(\frac{dI}{dV}\right)_{V=0}$ solving Equation (7) for R_{sh} yields:

$$R_{sh} = -\frac{\left(1+m\right)R_s}{m + \frac{q}{nkT_C} I_o\left(1 + R_s m\right)\exp\left(\frac{q}{nkT_C} I_{sc} R_s\right)} \quad (8)$$

The difficulty with Equation (8) is that, in addition to the short-circuit slope, it includes the parameters n and R_s which cannot be explicitly written in terms of the manufacturers' parameters. Thus, solving for these values simultaneously would involve a computationally intensive two-stage iteration scheme. It is proposed that Equation (8) may be simplified with the approximation:

$$R_{sh} = \frac{-1}{\left(\frac{dI}{dV}\right)_{V=0}} \tag{9}$$

To test the validity of Equation (9), several sample IV curves with different shunt resistances values were produced using the adapted five parameter model.

2.2. FORTRAN Component for I-V Characteristics

Several computer-based Software such TRNSYS [14], PV F-Chart [20], PVFORM [21], and Sandia National Laboratories [22] design handbooks methods exist for estimating the performance of PV systems. An important distinction among models is whether the model performs simulations over frequent, usually hourly, intervals or whether a reduced set of calculations is done based on average behavior over a longer, usually monthly interval. The detailed models require extensive weather data; Typical Meteorological Year (TMY) data. Another major distinction is whether or not the model is limited to use on maximum power-tracked systems. None of the previous available models are able to model direct-coupled loads of arbitrary I-V shape.

The method presented in this work uses long-term average weather data to estimate system output for directcoupled loads of any I-V shape. The equations in the previous sections are implemented to develop a numerical model to determine the characteristics of copper indium diselenide (CIS) module. The FORTRAN code necessitates three different groups of information: a set of parameters, an array of input variables and an array of output variables. The parameters are characteristic values which are fixed through out the run. Inputs to the code are the time dependent variables such as irradiance and temperature. The output depends on the parameters needed to analyze the performance of PV modules. The iteration process begins with an initial guess of V as an input to the PV array component. The PV array component calculates the corresponding current into the load component. The load component produces a new voltage serving as a second guess for PV array component. The successive substitution is continued until convergence is obtained and the operating point is achieved.

2.3. Non Linear Curve Fitting of Experimental Data

To extract the PV module parameters a non linear curve fitting program IVFIT of experimental I-V data proposed by Burgers et al. [19] is adapted for this study. Measurements of the I-V curves of solar modules are one of the most important means of obtaining information about PV modules. Straightforward least squares fitting of I-V curves leads to non accurate fit leading to a bad fit at the maximum power point and lower voltage values. The strongly varying slope makes it difficult to fit PV modules I-V curves with standard least square techniques. To deal with this problem, Burgers et al. [19] propose using weighting functions to minimize the area between data and fit instead of least squares procedure.

Voltage noise has a big influence on fitting due to the steep slope of an I-V curve for higher voltage values. So, Burgers et al. [19] have used Orthogonal Distance Regression (ODR) which is a mathematical method for fitting measurements and in the meantime it allows for computing the I-V curve parameters. The goodness of fit, c^2, is used to judge whether the oneor two-diode models are suitable. The computer code adapted employs an iterative scheme such that the fitted current and not the measured current are used to calculate the effect of series resistance. The program accounts for noise in the current as well as in the voltage signal of the measured datasets. ODR algorithms process both the error signal of the current and of the voltage close to I_{sc} and V_{oc}. Noise levels determined from the IVFIT is checked such that the IV curve fit at STC proved to be as accurate as possible.

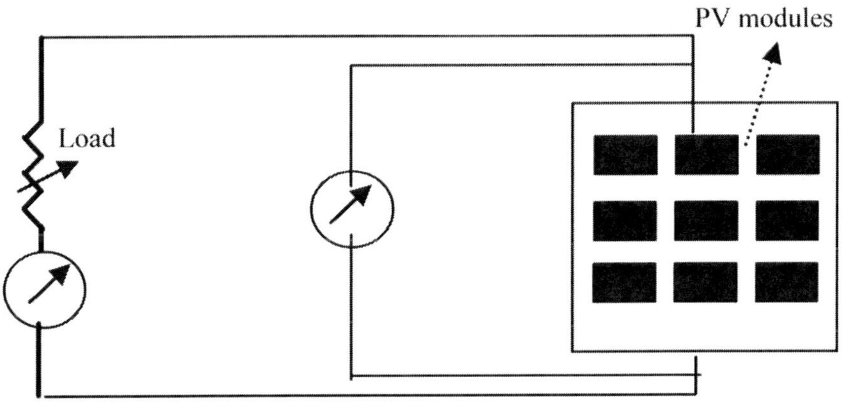

Figure 1. Schematic diagram of I-V setup for PV modules.

3. EXPERIMENTAL SETUP

Figure 1 shows a schematic diagram of the outdoor PV modules test facility installed at the College of Technological Studies, Kuwait. This test facility is designed to measure the I-V characteristics of different types of PV modules at different radiation and temperature levels in Kuwait climate. The solar modules are mounted on the roof of the main building in the College of Technological Studies facing south at a tilt angle 31° which is equal to Kuwait latitude. The place of the solar modules is chosen such that a shadow will not be cast into the solar modules at any time during the test period.

The test method proposed for the measurement of the I-V characteristics of PV modules at different radiation and temperature levels requires the measurements of the following parameters: 1) the global radiation on the solar module, 2) the I-V curve at different radiation levels and 3) the I-V curve at different temperature levels.

The intensity of the solar radiation incident on the module surface are measured and recorded by an Eppley Precision Spectral Pyranometers (PSP model) connected to the data acquisition system. Measurements are recorded for periods where variations in global radiation through the measurement period do not exceed ±20 W/m^2. The accuracy of the temperature measurements devices differ according to the purpose they are used for. The ambient air temperature should be measured within an accuracy of ±0.1°C. A standard resistance thermometer detector (RTD-PT100) is used to monitor the surrounding ambient temperature to guarantee high accuracy for this critical temperature. It is to be noted that the RTD sensor of the ambient temperature is shaded from direct and diffuse solar radiation.

The Keithley Source Meter instrument is adapted to perform I-V characteristics of different solar cell modules. The Keithley Source Meter instrument combines a programmable power source with the measurement functions of a digital multimeter. The Keithley LabTracer software allows users to coordinate the measurement and sourcing voltage or current and collect voltage and/or current readings, as well as a timestamp for each measurement set. One can sweep voltage or current allowing collection of complete test sequences. Results can be exported to windows based programs like EXCEL spreadsheet for calculation, and analysis.

Temperature measurements of PV devices are usually a difficult task. In general, measured temperatures should be those of the actual semiconductor junctions, but usually the only temperature that can be measured is at the rear surface of a cell. For modules, it not possible to contact individual cells and one is forced to apply temperature sensors to the rear surfaces. This leads to an error because the rear surface will normally be cooler than the cells laminated inside. At 1000 W/m^2, this error is about 2.5% for typical PV modules.

A five set of thermocouples are placed in the rear surface of the modules, one at each corner plus one at the module center to measure the module temperature. During the module tests, the corresponding values of environmental parameters were recorded by Keithley 2700 data acquisition device. All of this information is saved to different data files in a personal computer. The obtained measured data files are treated first to eliminate erroneous points. The I-V curves are then plotted and a numerical technique is adapted to analyze the measured data and extract the different modules parameters. The outdoor exposure tests were carried out during the period from 8th August 2009 to 10th August 2010. The sourcemeter is used to sweep voltage through the solar modules and measure current to obtain the I-V curve of the modules.

In PV system design, many of the calculations are based on the determination of module temperature from ambient temperature and NOCT [23-25]. From the linear regression of the plotting of the difference between module temperatures minus ambient temperature against the irradiation, a preliminary value of NOCT is obtained. This value is corrected to 800 W/m^2 and 20°C depending on the average values of ambient temperature and wind speed during the test. To obtain a good spread of measured points, simulations under different current ratings and various meteorological conditions were performed. The voltage and current of the module is recorded at each operating point with high precision digital numbers.

4. RESULTS AND DISCUSSIONS

The I-V curves of CIS thin-film modules were monitored for the period August 8, 2009 to August 10, 2010. the I-V curves are monitored using Keithley Sourcemeter. TheI-V characteristic of the PV modules while deployed outdoor represents the performance under real operating conditions Table 1 lists the manufacturer specifications of the module used in this study at reference conditions STC (G_{Ref} = 1000 W/m^2, $T_{C,Ref}$ = 25°C).

Table 1. Modules specifications.

Parameter	CIS Module
P (W)	10
I_{SC} (A)	0.72
V_{OC} (V)	24.0
I_{mp} (A)	0.58
V_{mp} (V)	17.0
μ_{ISC} (A/°C)	0.0026
μ_{VOC} (V/°C)	–0.10
Area (m^2)	0.108
Junctions	10

To check the reliability of the present numerical model, the measured I-V characteristics are compared to the corresponding characteristics obtained from the numerical model as shown in Figure 2. As illustrated by the figure, the two predictions agree very well. Also, the two predictions agree well with the IV curves provided by the module manufacturer. These facts indicate the reliability of the adapted numerical model. It should be noted that both temperature coefficients of short circuit current (m_{ISC}) and open circuit voltage (m_{VOC}) are provided by the manufacturer of the module. All the data are repeated at different days and time for the same radiation levels. So, the measured I-V points presented in the figures can be considered as the average points for the module during the test period. As predicted from Figure 2, there is a linear increase in photo-generated current with increased photon flux as irradiance levels increase, resulting in the observed increase in current with increased irradiance.

The variation of power and the values of P_{max} produced by the module is calculated for each radiation level using the present developed model. These values are compared with the ones extracted using the IV fit program [19] and are found to be in good agreement. It was found that there is a significant increase in the module power with increased irradiance levels. It should be mentioned that the maximum power point is not always corresponding to solar noon which is 12:00. The reason for this may be attributed to the fact that the amount of irradiance received by the module is at its maximum at solar noon. Hence I_{SC}, is also at its maximum. In the meantime there is a drop in open circuit voltage, V_{OC}, around solar noon, which may be attributed to the elevated temperatures at this time of the day Also, it is observed that the voltage corresponding to maximum power point changes with radiation levels. The operating temperature of the PV module is one of the most important parameters affecting the solar module power output and its efficiency to convert sunlight into electricity. The output of the photovoltaic module de creases with increase of the cell temperature. In order to improve PV module maximum efficiency and output power, the module temperature must be reduced.

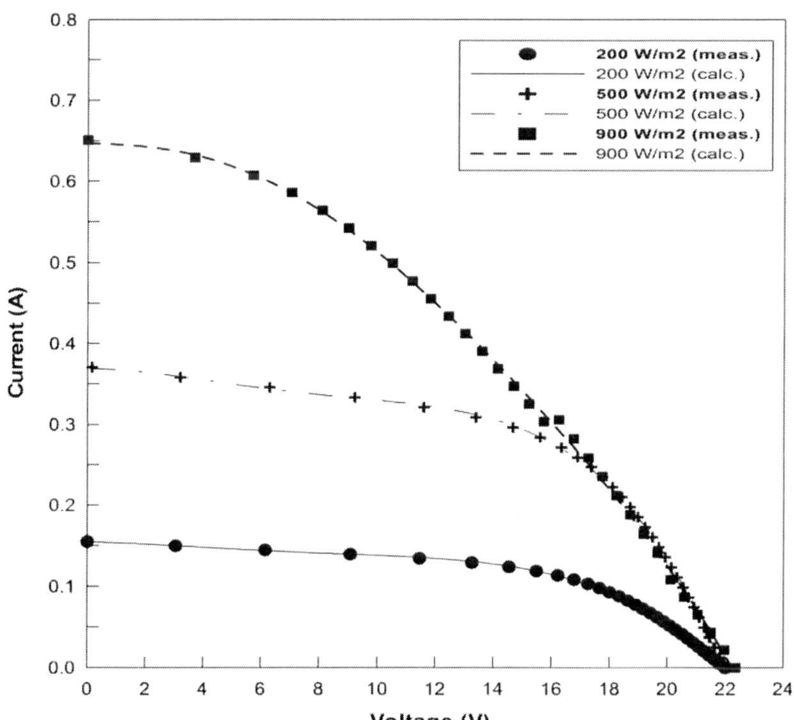

Figure 2. Comparison between measured and predicted. I-V characteristics for CIS module (Ta = 20°).

The effect of temperature on the I-V curve of CIS module is illustrated by Figure 3. Different curves are measured around noon on different days at different module temperatures. The presented graph represents the average measurements for different days and times but at the same radiation intensity level (800 W/m^2). The decrease in V_{oc} with increase in temperature is clearly illustrated in the figures. Also, the figure clearly indicates that the open circuit voltage, V_{oc}, is more temperature dependent than the short circuit current I_{sc}. The largest drop in voltage due to about 20°C increase in temperature is approximately 8.8% drop in V_{oc} which is not compensated for by the relatively small increase in current, (2.9% in I_{sc}), resulting in a reduction in P_{max} of 6.3%. If not taken into account, excessive temperatures coupled with variation in irradiation, can lead to under-designing of PV systems, which in turn may lead to system failure. As stated before, the comparison between the results obtained from the numerical model and the experimental I-V data indicated the accuracy of the present measurements.

So, the non linear curve fitting technique proposed by Burgers et al. [19] is adapted to extract the PV module parameters from the experimental I-V data at different environmental conditions. The five parameters of the PV module are: I_L, I_o, R_s, R_{sh} and ideality factor, n, which is a measure of cellular imperfection. The I_L for CIS module is observed to depend strongly on the solar irradiance. In addition, results imply that I_L directly proportional to irradiance intensity level. In the following we present a discussion for the effect of environmental conditions on the other four parameters of CIS module.

The module parameters variation with radiation at temperature 20°C is presented in Figure 4. As indicated by the figure, R_{sh} increases with radiation at low radiation values (<400 W/m^2). The increase in R_{sh} with radiation at low radiation values may be attributed to the existence of local inhomogeneities leading to non uniform current flow or to the charge leakage across the p-n junction in the cell. The reason of this behavior is that R_{sh} is associated with the localized defect regions which in turn have a larger concentration of traps that make them electrically active. We believe that the electrical activity of traps in this case is stronger at low irradiance. The same behavior has been observed by Priyanka et al. [26] and Cunningham et al. [27] for silicon solar cells. Contrary to their results, as radiation increases at high radiation val ues (>400 W/m^2), R_{sh} begins to decease sharply. As the radiation intensity increases, the traps start getting filled and this of course results in increase of leakage current which in turn causes the sharp decrease observed in R_{sh}. As the irradiance increases up to the value 900 W/m^2, the leakage current starts to increase more and more and R_{sh} reaches very low values as indicated by Van Dyk et al. [28] in their investigation of the effect of increasing irradiance intensity on R_{sh} for CIS modules. Also, at high values of radiation, some other effects as the heating of the cell can degrade R_{sh} to these low values [29].

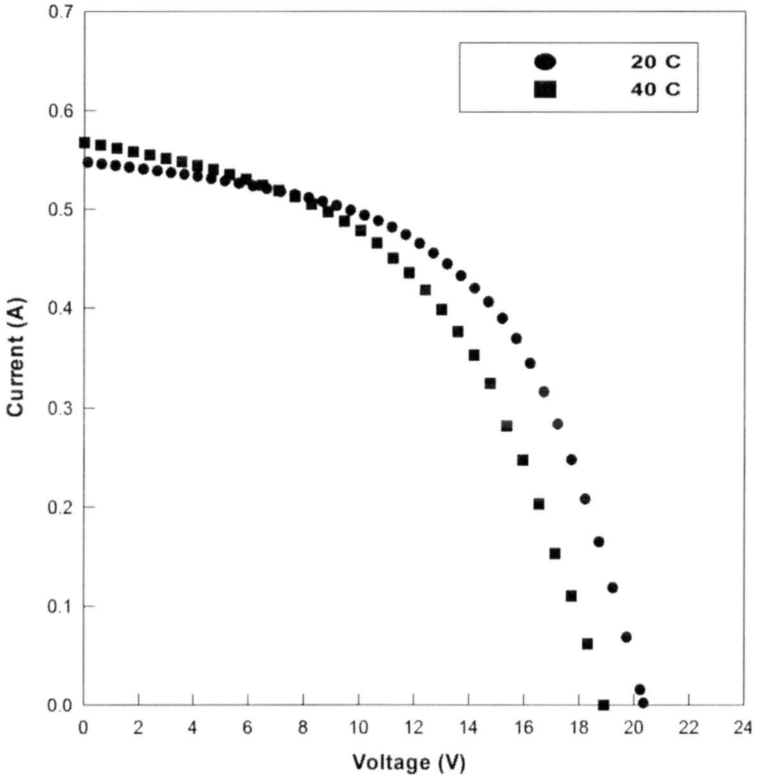

Figure 3. Variation of I-V characteristics with temperature for CIS module (G = 800 W/m^2).

It is known that R_{sh} is due to shunt paths that lead the current away from the intended load and their effects are detrimental to the module performance especially at low irradiance. The number of shunts in thin film cell may increase after prolonged exposure to light. The increase in the number of shunts increases the effect of shunt currents in the module and hence decreases R_{sh} with increasing irradiance level. So, the sharp decrease of R_{sh} for high radiation level (>400 W/m^2) indicates that some of the module cells increase shunt currents and conesquently stop contributing to the total output of the module. R_S represents the total resistance that current meets when crossing the cell from one end to the other, so it represents resistances in the cell solder bonds, emitter and base region, cell metallization, cell interconnect bus bars and resistances in the junction books termination It is evident that R_S decreases the value of V_{OC}, I_{SC} and FF. As concluded previously [28,30], as the light intensity incident on the module increases, R_S and the applied voltage decrease. When the irradiance intensity increases, R_S decreases but with a very slow rate at the two studied temperatures (20°C and 40°C) as presented in Figures 4 and 5.

This decrease may be attributed to an increase in the conductivity of the active layer with the increase of the radiation intensity. The slow rate decrease of R_S has little effect on solar module performance compared with the significant change of other module parameters.

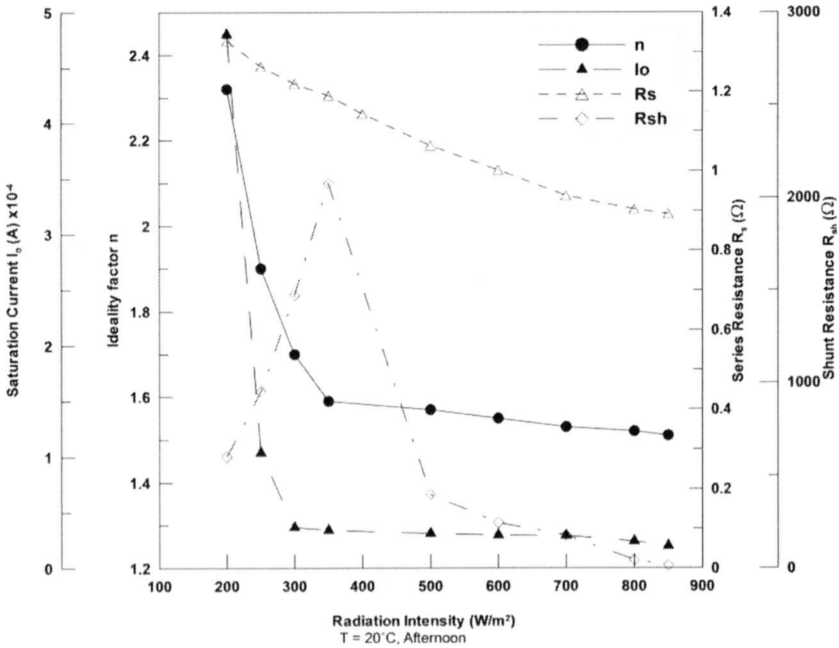

Figure 4. The module parameters variation with radiation intensity at module temperature 20°C.

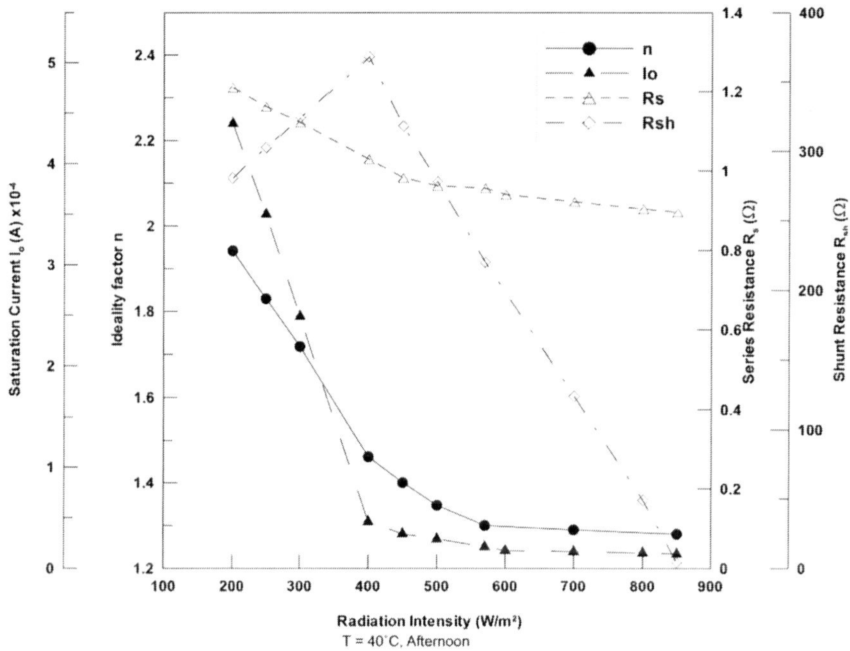

Figure 5. The module parameters variation with radiation intensity at module temperature 40°C.

Also Figures 4 and 5 shows the effect of irradiance level on the ideality factor n and the saturation current Io at 20°C and 40°C respectively. It is well known that CIS materials are highly compensated with fluctuations in stoichiometry and composition which cause electrostatic fluctuations and band gap fluctuations respectively [31]. In case of polycrystalline thin films, it is almost trivial that not only electrostatic potential fluctuations at grain boundaries must occur, but also band gap fluctuations within the grain and from grain to grain. The primary effect of grain boundaries on polycrystalline CIS results from the interruption of the long range order provided by the periodicity of the crystal. Such interruption leads to defect electronic states in the energy gap at its center and at band edges, which trap the photo-generated carriers causing recombination. For CIS solar cells and modules the values of n and I_o are due to different types of recombination of photo-generated carriers or minority carriers [32]. Recombination losses in CIS don't affect the I_{SC} only but also the V_{OC} and the FF. Ideality factor equals to 2 is valid for recombination with the states exactly in the gap centre, while ideality factor equals to 1 is valid for recombination at the band edges [33]. The saturation current I_o has a dramatic and direct effect on V_{oc} as directly predicted from the following equation:

$$V_{oc} = \frac{nkT}{q} \ln\left[\frac{I_{ph} - I_o}{I_o}\right] \approx \frac{nkT}{q} \ln\left[\frac{I_{ph}}{I_o}\right] \quad (10)$$

Figures 4 and 5 show that as irradiance intensity increases, n and I_o decrease first sharply in the low range of radiation intensity (<400 W/m^2) and this increase becomes small for irradiance values greater than 400 W/m^2. This behavior is attributed to the non-radiative recombination in the space charge region which is dominant for CIS devices and this type of recombination is more effecttive at low radiation intensity and low junction voltage [34]. With increasing irradiance intensity the contribution of non-radiative space charge recombination to the total recombination processes start to decrease as n and I_o values in this case are mainly due to the bulk and surface recombination mechanisms. The module parameters variation with radiation at module temperature 40°C is presented in Figure 5. The previous observations and conclusions regarding the module parameters R_{sh}, R_s, n and I_o obtained at 20°C are applicable again at 40°C. However, it is noticed that when the irradiance intensity increases, R_{sh} increases for irradiance values less than 400 W/m^2 or decreases for irradiance values greater than 400 W/m^2 but the rate of change at 40°C is less than that at 20°C. Also, there is a great difference between the peak values of R_{sh} at both temperature levels. The maximum value of R_{sh} at 20°C is about 2100 W compared to about 475 W at 40°C.

The values of the CIS module parameters shown in Figures 4 and 5 along with I_{sc} are implemented in equation (4) to calculate the module performance parameters, i.e., V_{oc}, FF and h. It should be mentioned that the module performance parameters obtained from IVFIT software compare well with the corresponding values obtained from the present theoretical predictions. Figures 6 and 7 show the variation of V_{oc}, FF and module efficiency (h) with radiation intensity for temperatures 20°C and 40°C for the same day of measurements of Figures 4 and 5. The module efficiency h given by Equation (3) can be rewritten in the following form:

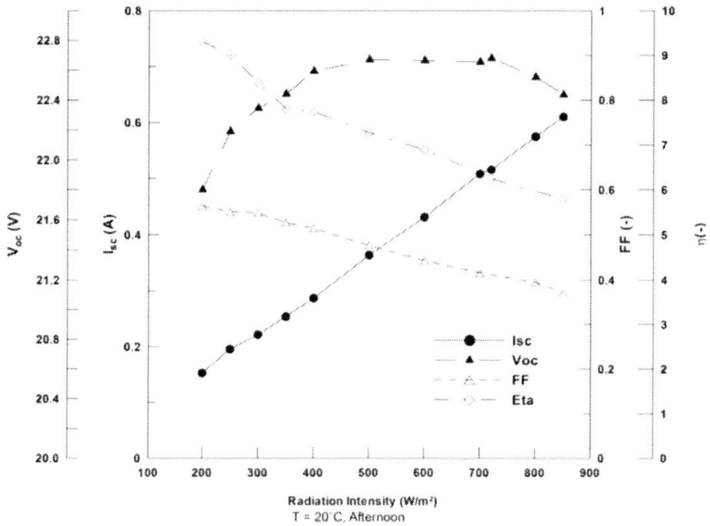

Figure 6. Variation of module performance parameters with radiation intensity at module temperature 20°C.

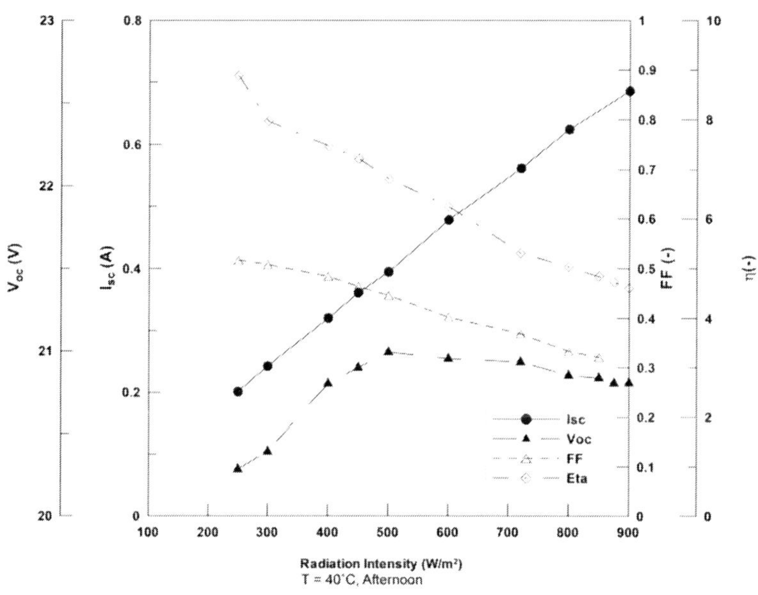

Figure 7. Variation of module performance parameters with radiation intensity at module temperature 40°C.

$$\eta = \frac{FFI_{sc}V_{oc}}{GA} \tag{11}$$

As predicted from Equation (11), I_{sc} is linearly proportional to the irradiance, so it can be considered that it has no obvious effect on efficiency as the increase in I_{sc} is compensated by the increase in radiation intensity. So, the decrease in efficiency with increasing irradiance level can be attributed to the combined effect of both V_{oc} and FF. Fill factor continuously decreases at small rate with increasing irradiance level similar to the behaviour of n and I_o previously discussed. On the other hand, V_{oc} increases with irradiance intensity up to the value 400 W/m^2 then decreases by a very slow rate also similar to the behaviour of n, I_o and R_{sh}. So, as a result the efficiency h decreases continuously with irradiance level due to the combination effect of V_{oc} and FF. The same behaviour is depicted again at high temperature level (40°C) as illustrated by Figure 7. It is worth to mention that the effect of increasing temperature from 20°C to 40°C results in decrease in the value of V_{oc} by about 8.8%. It should be mentioned that the order of precision for the predicted values of the CIS parmeters is in the range of accepted errors. Estimated errors in the present measurements are found to fluctuate somewhere around 2%, a value which is relatively reasonable compared to other published works.

5. CONCLUSIONS

An experimental setup is designed at the College of Technological Studies, Kuwait to monitor the I-V characteristics of PV modules under realistic outdoor conditions in the period from August 8, 2009 to August 10, 2010. The data obtained showed the value of monitoring PV modules over a period of time. The monitoring also yielded useful results on how the PV module performed over an extended period of time with regard to energy generating capability and operational efficiencies. Measurements are carried out to determine the I-V characteristics of copper indium diselenide (CIS) module at different radiation and temperature ranges. A theoretical model is developed to calculate the module parameters and module performance parameters. The validation of the used numerical model has been checked and results proved the reliability of the present developed numerical model. This work successfully emphasizes the importance of outdoor performance monitoring as there was a difference between outdoor measurements and indoor measurements published by the module manufacturer. Running the program code several times has allowed reaching the convergence of the equations system; then the determination of the main parameters of CIS module. Based on the results obtained, the following conclusions are drawn:

- Photo-generated current linearly increase with increased irradiance.
- A comparison between the characteristics given by the developed numerical model and those obtained by outdoor measurement tests demonstrated the reliability of the developed numerical model.
- The largest drop in V_{oc} due to about 20°C increase in temperature is approximately 8.8% which is not compensated for by the relatively small increase in current, (2.9% in I_{sc}), resulting in a reduction in P_{max} of about 6.3%.
- The shunt resistance R_{sh} increases with radiation at low radiation values (<400 W/m^2). While it begins to decease sharply as radiation increases at high radiation values greater than 400 W/m^2.
- The series resistance decreases as the irradiance intensity increases, but with a very low rate at the two studied temperatures (20°C and 40°C). So, R_s has little effect on module performance in comparison with the significant effect of other module parameters.
- As irradiance intensity increases, ideality factor n and saturation current I_o decrease first sharply in the low range of radiation intensity (<400 W/m^2) and this decrease becomes smaller for irradiance values greater than 400 W/m^2.
- There is a great difference between the peak values of R_{sh} at both studied temperature levels. The maximum value of R_{sh} at 20°C is about 2100 W compared to 475 W at 40°C.
- Fill factor FF continuously decreases at small rate with increasing irradiance level similar to the behaveior of n and I_o. Open circuit voltage V_{oc} increases with irradiance intensity up to 400 W/m^2 then decreases by a very small rate similar to the behavior of n, I_o and R_{sh}.
- The efficiency h decreases continuously with irradiance level due to the combined effect of V_{oc} and FF for both studied temperatures.
- The present work did not consider the effect of variation of spectral irradiance through the day on the module parameters. This effect should be studied to accurately predict the module performance.
- The order of precision for the predicted values of the CIS parameters is in the range of accepted errors. The estimated errors fluctuate somewhere around 2%, a value which is relatively reasonable compared to previously published works.

6. ACKNOWLEDGEMENTS

The research team would like to express their sincere gratitude to The Public Authority for Applied Education and Training (PAAET) for financial support of this work under the project No. (TS-08-02).

REFERENCES

1. M. Wolf and H. Rauschenbach, "Series Resistance Effects on Solar Cells Measurements," Advanced Energy Conversion, Vol. 3, No. 2, 1963, pp.455-479.
2. K. Rajkanan and J. Shewchun, "A Better Approach to the Evaluation of the Series Resistance of Solar Cells," SolidState Electron, Vol. 22, No. 2, 1979, pp.193-197.
3. S. K. Agarwal, R. Muralidharan, Amit Agarwal, V. K. Tiwary and S. C. Jain, "A New Method for the Measurement of Series Resistance of Solar Cells," Journal of Physics D: Applied Physics, Vol. 14, No. 9, 1981, pp. 1643-1646.
4. G. L. Araujo and E. Sanchez, "A New Method for Experimental Determination of the Series Resistance of a Solar Cell," IEEE Transactions on Electron Devices, Vol. 29, No. 10, 1982, pp. 1511-1513.
5. M. Chegaar, Z. Ouennouchi and A. Hoffmann, "A New Method for Evaluating Illuminated Solar Cell Parameters," Solid-State Electron, Vol. 45, No. 2, 2001, pp. 293- 296.
6. E. E. van Dyk, E. L. Meyer, F. J. Foster and A. W. R. Leitch, "Long-Term Monitoring of Photovoltaic Devices," Renewable Energy, Vol. 25, No. 2, 2002, pp. 183- 197.
7. E. Radziemska, "Dark I-U-T Measurements of Single Crystalline Silicon Solar Cells," Energy Conversion Management, Vol. 46, 2005, pp. 1485-1494.
8. M. Priyanka and S. N. Singh, "A New Method for the Measurement of Series and Shunt Resistance of Silicon Solar Cells," Solar Energy Materials and Solar Cells, Vol. 46, No. 9-10, 2007, pp. 137-142.
9. K. I. Ishibashi, Y. Kimura and M. Niwano, "An Extensively Valid and Stable Method for Derivation of All Parameters of a Solar Cell from a

Single Current-Voltage Characteristics," Journal of Applied Physics, Vol. 103, No. 9, 2008, pp. 094507-094513.

10. G. P. Smestad et al., "Reporting Solar Cell Efficiencies in Solar Energy Materials and Solar Cells," Solar Energy Materials and Solar Cells, Vol. 92, No. 4, 2008, pp. 371- 373.
11. S. Mohammed and M. Ahmed, "Outdoor Testing of Photovoltaic Arrays in the Saharan Region," Renewable Energy, Vol. 33, No. 12, 2008, pp. 2516-2524.
12. S. N. Singh and M. Husain, "Effect of Illumination Intensity on Cell Parameters of a Silicon Solar Cell," Solar Energy Materials and Solar Cells, Vol. 94, No. 9, 2010, pp. 1473-1476.
13. T. U. Townsend, "A Method for Estimating the LongTerm Performance of Direct-Coupled Photovoltaic System," Mechanical Engineering, University of Wisconsin-Madison, Madison, 1989.
14. S. A. Klein et al., "TRNSYS Users Manual," Version 14.1, University of Wisconsin-Madison, Engineering Experimental Station, 1994.
15. A. Ortiz-Conde, F. J. G. Sanchez and J. Muci, "New Methods to Extract Model Parameters of Solar Cells from the Explicit Analytic Solutions of Their Illuminated I-V Characteristics," Solar Energy Materials and Solar Cells, Vol. 90, 2006, pp. 352-361.
16. W. De Soto, S. A. Klein and W. A. Beckman, "Improvement and Validation of a Model for Photovoltaic Array Performance," Solar Energy, Vol. 80, No. 1, 2006, pp. 78-88.
17. F. Almonacid, C. Rus, L. Hontoria, M. Fuentes and G. Nofuentes, "Characterisation of Si-Crystalline PV Modules by Artificial Neural Networks," Renewable Energy, Vol. 34, No. 4, 2009, pp. 941-949.
18. K. Ishaque, Z. Salam and H. Taheri, "Simple and Accurate Two-Diode Model for Photovoltaic Modules," Solar Energy Materials and Solar Cells, Vol. 95, No. 2, 2011, pp. 586-594.
19. A. R. Burgers, J. A. Eikelboom, A. Schonecker and W. C. Sinke, "Improved Treatment of the Strongly Varying Slope in Fitting Solar Cell I-V Curves," Proceedings 25 the IEEE PVSC Conference, Washington DC, 1996, pp. 569-572.
20. S. A. Klein and W. A. Beckman, "PV f-Chart User's Manual: f-Chart Software," Middleton, El Paso, 1983.
21. D. F. Menicucci and J. P. Fernandez, "User's Manual for PVFORM," 1988.

22. Stand-Alone Photovoltaic Systems, "A Handbook of Recommended Design Practices," Sandia National Laboratories, Albuquerque, SAND87-7023, 1988.

23. M. C. A. Garcia and J. L. Balenzategui, "Estimation of Photovoltaic Module Yearly Temperature and Performance Based on Nominal Operation Cell Temperature Calculations," Renewable Energy, Vol. 29, No. 12, 2004, pp. 1997-2010.

24. D. King, J. Kratochvil and W. Boyson, "Temperature Coefficients for PV Modules and Arrays: Measurement Methods, Difficulties, and Results," 26th IEEE PV Specialists Conference, Anaheim, 29 September - 3 October 1997, pp. 1183-1186.

25. E. Skoplaki and J. Lyvos, "Operating Temperature of Photovoltaic Modules: A Survey of Pertinent Correlations," Renewable Energy, Vol. 34, No. 1, 2009, pp. 23- 29.

26. M. Priyanka and S. N. Singh, "A New Method for the Measurement of Series and Shunt Resistance of Silicon Solar Cells," Solar Energy Materials and Solar Cells, Vol. 91, No. 2-3, 2007, pp. 137-142.

27. D. W. Cunningham, A. Parr, J. Posbic and B. Poulin, "Performance Comparison between BP Solar Mono and Traditional Multicrystalline Modules," 23rd European PV Solar Energy Conference Proceedings, Valencia, 1-5 September 2008, pp. 2829-2833.

28. E. E. Dyk and E. L. Meyer, "Analysis of the Effect of Parasitic Resistances on the Performance of Photovoltaic Modules," Renewable Energy, Vol. 29, No. 3, 2004, pp. 333-344.

29. M. Priyanka, S. N. Singh and M. Husain, "Temperature Dependence of I-V Characteristics and Performance Parameters of Silicon Solar Cells," Solar Energy Materials and Solar Cells, Vol. 92, No. 12, 2008, pp. 1611-1616.

30. V. Perraki and A. Gianniou, "The Effect of the Series Resistance on the Photovoltaic Properties of CIS Solar Cells," 25th EUPVSEC, Valencia, 6-10 Septemper, 2010.

31. J. H. Werner, J. Mattheis and U. Rau, "Effeciency Limitations of Polycrystalline Thin Film Solar Cells: Case of (In, Ga)Se_2," Thin Solid Films, Vol. 480-481, No. 1, 2005, pp. 399-409.

32. U. Rau, K. Taretto and S. Siebentritt, "Grain Boundaries in Cu (In, Ga) (Se, S) 2 Thin Filmsilar Cells," Applied Physics A, Vol. 96, No. 1, 2009, 221-2234.

33. S. Siebentritt, "What Limits the Efficiency of Chalcopyrite Solar Cells?" Solar Energy Materials and Solar Cells, Vol. 95, No. 6, 2011, pp. 1471-1476.

34. F. Khan, S. N. Singh and M. Husain, "Effect of Illumination Intensity on Cell Parameters of a Silicon Solar Cell," Solar Energy Materials and Solar Cells, Vol. 94, No. 9, 2010, pp. 1473-1476.

INDEX